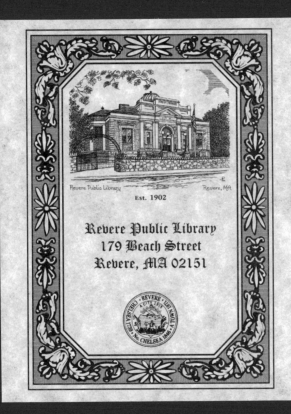

SCIENCE UNSHACKLED

SCIENCE
UNSHACKLED

How Obscure, Abstract, Seemingly Useless
Scientific Research Turned Out to Be the Basis
for Modern Life

C. RENÉE JAMES

Johns Hopkins University Press

Baltimore

© 2014 Johns Hopkins University Press
All rights reserved. Published 2014
Printed in the United States of America on acid-free paper
9 8 7 6 5 4 3 2 1

Johns Hopkins University Press
2715 North Charles Street
Baltimore, Maryland 21218-4363
www.press.jhu.edu

Library of Congress Cataloging-in-Publication Data

James, C. Renée, author.
    Science unshackled : how obscure, abstract, seemingly useless
scientific research turned out to be the basis for modern life /
by C. Renée James.
        pages   cm
    Includes bibliographical references and index.
    ISBN 978-1-4214-1500-0 (hardcover : alk. paper) —
ISBN 1-4214-1500-3 (hardcover : alk. paper) —
ISBN 978-1-4214-1501-7 (electronic) — ISBN 978-1-4214-1501-1
(electronic) 1. Science—Miscellanea. 2. Research—Miscellanea.
I. Title.
    Q173.J34 2014
    507.2—dc23   2014006747

A catalog record for this book is available from the British Library.

Special discounts are available for bulk purchases of this book.
For more information, please contact Special Sales at 410-516-6936
or specialsales@press.jhu.edu.

Johns Hopkins University Press uses environmentally friendly
book materials, including recycled text paper that is composed
of at least 30 percent post-consumer waste, whenever possible.

For my amazing children
Sean, Megan, and Jamie,
who have taken me on countless unexpected random walks
of discovery

CONTENTS

PREFACE

The ideas in this book started to gel during the first full week of July 2011. It was during this week that I watched the last launch of NASA's Space Shuttle program. Unfortunately I saw it from nearly a thousand miles away, after having told myself for 30 years that I would get to Florida to see a shuttle launch "someday." Tears filled my eyes on that summer morning when I realized that all those "somedays" had come and gone.

That same week I also watched what appeared to be the end of the James Webb Space Telescope, the new and improved successor to the Hubble Space Telescope. Having gone billions of dollars over budget and several years behind schedule, the JWST was declared by the United States Congress to be a pointless money pit, making it an easy target during the worst economic period since the Great Depression. Without warning, Congress announced that the entire project was canceled. Baffled, I wondered why our elected officials, who were supposed to represent the viewpoints and priorities of their constituents, would cancel a project that had already been deemed important enough to construct.

Naturally, the community of astronomers was floored. Within hours, the largest professional astronomer society on the planet, the American Astronomical Society, had issued a strong statement against cutting the JWST program. They pointed out (rightly) that billions had already been spent, that cost overruns and unforeseen delays come with the territory of developing cutting-edge technology (indeed that delays necessarily create cost overruns), and that the telescope has the potential to revolutionize astronomy more than Hubble did.

This is a bold claim. Audacious, even. JWST could outdo Hubble? Most assuredly, claims Senator Barbara Mikulski, the chair of the Senate Appropriations Committee that funds NASA. "Going from the Hubble Space Telescope to the James Webb Space Telescope is like going from a biplane to the jet engine," she stated at a 2014 press conference about progress on the new telescope.

Besting the Hubble Space Telescope is no small matter. As pointed out dramatically in David Gaynes's moving 2012 documentary *Saving Hubble*, the Hubble Space Telescope is one of the few scientific instruments that the general public can identify by name, its moniker and stunning images trickling quietly into popular culture for more than two decades. Over a third of the population has never known life without a visible-light telescope in space.

As for the JWST, the astronomy community was in complete agreement with Congress that we should naturally address any mismanagement of JWST, just as we should with any publicly funded project. But please . . . let the telescope see first light. Let future generations experience the awe that Hubble has brought to this one. Fortunately, the combined professional and public outcry seemed to change the minds of the lawmakers, who restored funding by the end of 2011 but left plenty of uncertainty in the telescope's ultimate fate.

As life changing as that week was for me and other astronomy aficionados, however, not everyone was so moved. Beneath an Internet news story on the Space Shuttle's swan song was a stunningly bitter comment: "What do they plan on accomplishing? Can they cure cancer up there?"

The tone was even uglier for the JWST: "Stop wasting money on these toys," they effectively said. "And make those brainy astronomers work on real-life problems staring them in the face rather than devote time and energy and money staring at some impossibly distant galaxy." As if that weren't discouraging enough, NASA's next mission spawned another litany of negative comments. Despite the successful and awe-inspiring landing of the Mars *Curiosity* rover during the summer of 2012, many felt it was more money wasted on pointless science research.

Every time I saw these sentiments, my brain rebelled. "Surely everyone realizes how important science is!"

Then my brain quietly and tentatively added, "Don't they?"

Clearly not. Too many people seem not to realize that individual research projects are usually just a pixel in a much larger picture. What has become painfully clear from those comments is that too few people seem to appreciate that those apparently pointless, small research projects can sometimes become the surprising life changers, just as the frumpy-looking underdog in a talent contest can bring down the house.

Hence the creation of this book, which presents the sometimes messy histories behind research projects that have had awesome unintended benefits. Although I'm a professional astronomer, I knew right away that my net had to be cast as widely as possible. Beyond JWST. Beyond astronomy. Science as a whole is under the suspicious eye of the societal microscope, and it was time to try to do something about it.

But with committee meetings and proposal deadlines and classes and a family, I let the book slide into second, then third, then fourth place, then nth place in my priorities. It progressed in fits and starts, chugging slowly along much like the JWST: it was being worked on, but it was merely creeping closer to launch. Comments I read posted beneath science news articles remained as jaded as ever, and I watched and cheered as other people valiantly took up the mantle to fight for pure science research while I prepared a committee report.

Then one day—April 23, 2013, to be exact—I read that Congressman Lamar Smith from my home state of Texas had urged the National Science Foundation to require grant applications to explain how proposed research will directly benefit the American people. While not an entirely new requirement (scientists have long had to provide a statement of "broader impacts"), it seemed to galvanize my resolve, and I wondered, "How would a proposal to do a radio-telescope search for exploding black holes fare in today's climate?"

I was pretty sure I knew the answer to that one. I also knew that if I wanted to try to show that society has misplaced its priorities, it was time to reprioritize my own life and put the book back at the top of the list. It was time to demonstrate that allowing (and publicly funding) scientists to explore the natural universe simply for the sake of satisfying their curiosity has improved our lives immeasurably.

But that is not all. It was also time to show that humanity has learned the same lesson repeatedly: derided "impractical" research projects have often led to unintended consequences that have changed how you live your life in ways you might never imagine.

ACKNOWLEDGMENTS

When this idea first began to nag at my brain soon after the last Space Shuttle launch, I mentioned it to the kind editors at Johns Hopkins University Press. After a few iterations on exactly which topics were to be explored (and how), Vincent Burke and I converged on the format you see here.

The concept also resonated with others. *Astronomy* magazine editor David Eicher accepted a very abbreviated, astronomy-only capsule summary of a few of the findings discussed in this book. A small portion of the stories found in parts I, III, and V was published in the May 2012 issue of *Astronomy* and is reprinted with permission here. In addition, some of the background on the history of understanding the composition of the Sun (part V) was previously published in the January/February 2005 *Mercury* magazine and is reprinted with the kind permission of *Mercury* editor Paul Deans.

Scientists across the globe enthusiastically related their own tales of the benefits of pure science, the frustrations with government cutbacks and requirements for "practically applicable science, only" projects. For instance, Ronald Hoy at Cornell University, who admits simply to being interested in the sounds of bugs, graciously answered every question I sent his way. Ultimately, the part that I'd intended to craft around a fly's contribution to the hearing aid industry was arrested in its larval phase, but I wish to thank Dr. Hoy for his gracious and thoughtful responses, many of which made their way into the introductory material.

Writing, like science, is best done with the input of people with a wide variety of skills. Physicist Joel Walker (Sam Houston State

University) provided very helpful analogies to explain the principles of relativity found in part I. Forensic scientist David Gangitano (Sam Houston State University) gave me a personal tour of his DNA sequencing lab, leaving me mystified at the precision of the process that could be carried out in something that appeared for all the world like a mass-produced bread maker. Discussions with him could fill their own book, and I know I failed to do the field of DNA forensics any justice in the brief introduction found in part II. I also appreciate the generous help of biologists who tried to retrain me in the basics of the history of genetics, particularly Anne Gaillard (Sam Houston State University), who read multiple drafts of part II, never failing to comment promptly and thoroughly.

Ron Ekers and John O'Sullivan (both at Australia's CSIRO), two-thirds of the trio that went on the wild goose chase of exploding black hole hunting, provided much needed clarification on many aspects of black holes, radio astronomy, the instrumentation that led to WiFi, and the benefits of science to society. Dr. Ekers also read and commented extensively on part III in its entirety, and for that I am in his debt.

Baldomero "Toto" Olivera, whose work has appeared in dozens of popular science articles and YouTube lectures, was amazingly generous with his time, reviewing part IV, which discusses how his curiosity about cone snails has opened up a whole medicine cabinet of treatments. If I had the chance to rewind and explore a different scientific career, I would probably knock on Dr. Olivera's door.

On the cusp of turning a seemingly impractical research project into a cancer-treating life changer at Ohio State University are Sultana Nahar and Anil Pradhan, both of whom were gracious enough to explain the more esoteric aspects of x-ray astrophysics and atomic physics that are featured in part V. Their encouragement at many stages of the manuscript's progress was very much appreciated, and I am eager to see what becomes of their novel approach to battling tumors.

There are plenty of people whose work wasn't featured in the book, yet who were invaluable in its preparation. Derek Wills of the

University of Texas at Austin helped verify astronomy and physics content and had a stunningly keen eye for stylistic issues and
grammatical errors in all five parts of the book. Tim Slater (University of Wyoming), who has been around the book-writing block a
few times, helped me bring some of the text down to Earth for the
nonastronomer. He provided an abundance of stylistic suggestions
that I eagerly incorporated. David Toback of Texas A&M University
pointed out some historical inaccuracies and provided both encouragement and helpful feedback. Rob Thacker of St. Mary's University
in Halifax, Nova Scotia, Canada, directed me to useful references
regarding the socioeconomic impact of pure research science. Jeff
Foust, editor of the *Space Review*, was generous enough to provide the
full audio recording of Steven Weinberg's 2011 public talk addressed
the problem of science funding and the fate of large-scale projects
like the James Webb Space Telescope. Finally, Harold Zakon of the
University of Texas at Austin is partially to blame for this entire book,
as he authored a newspaper editorial that gave the tiniest taste of the
benefits of pure, curiosity-driven research. He also suggested possible
directions that the parts of the book might take long before the first
words were set down.

Once the words were largely crafted, my husband, Sam Beard, uncovered a number of places where I assumed too much prior science
knowledge from the reader. High school student and astronomy enthusiast Xzavier Flowers of Lancaster, Texas, did the same, and he
provided several helpful comments. I hope the science culture he in-
herits encourages and rewards unbridled curiosity. Once I had settled
on most of the words in this book, it took the keen eye of Michele
Callaghan at Johns Hopkins University Press to further refine them
(and to remove the seventeen instances of the same word in one of
the chapters).

None of this could have been accomplished were it not for the
support of Sam Houston State University. Specifically, Provost Jaimie
Hebert not only introduced me to *The Music of the Primes*, practically
the epitome of pointless-curiosity-turned-practical but also whole-
heartedly encouraged this project, despite its departure from typical

university research. I have always been grateful that faculty members at SHSU—and indeed at universities in general—are free to explore ideas that catch their attention, as these are the ideas that have the potential to change lives.

This was a project that almost invariably started lively discussions, whether in Facebook groups, by e-mail, or in person, and I'm bound to be forgetting a number of key players who gave me ideas and helped keep this project moving. For those omissions, I apologize.

Finally, I want to thank my family for trusting that whatever I was doing was worth the piles of paper on the dining room table and the weeks of time staring at a screen instead of doing fun things with them, particularly in the months leading up to my oldest son's year-long stint in Germany. They patiently endured videos of cone snails eating clown fish that looked like Nemo (which still haunt Jamie, my youngest), forced readings of paragraphs I was particularly proud of, animated discussions of the importance of curiosity-driven science when some Internet article suggested otherwise, and a thousand other little things that I subjected them to during the course of writing this book. I think all of them could now provide good arguments for funding pure science research at this point. I only hope that more voters do.

SCIENCE UNSHACKLED

INTRODUCTION

When I woke up on September 28, 1928, I certainly didn't plan
to revolutionize all of medicine by discovering the world's first
antibiotic.

Alexander Fleming, discoverer of penicillin

What is the exact chemical composition of the Sun? How
do parasitic flies zero in on their desired host crickets? Is it possible
to make a lab culture of this strange pink organism found in hot
springs? Do tiny black holes really exist and explode? What makes
some jellyfish glow in the dark? How can we figure out whether matter really does warp space-time?

But more important, for what useful purpose do birds sing?

Johannes Kepler responded to the last question in his book *Myste-
rium Cosmographicum* (*The Cosmic Mystery*) more than 400 years ago.
"We do not ask for what useful purpose the birds do sing, for song is
their pleasure," he wrote. "Similarly we ought not ask why the human mind troubles to fathom the secrets of the heavens."

"Fine, but fathom the secrets of the heavens on someone else's
dime" has become the modern response. Amid a flat, seemingly
hopeless economy, dissatisfaction with what is seen as wasteful government spending is increasing. And, judging by the titles of their
federally funded grant proposals, huge numbers of scientists are involved in endeavors that are just too far removed from everyday life
to be anything but frivolous.

"Shrimp on treadmills? My hard-earned money funded a study of
shrimp on treadmills?!" Although treadmill-bound shrimp made for

a clever set of YouTube videos, many people were more than irritated that half a million dollars was spent apparently amusing a scientist. It was the only real press that this study received. The scientists involved vocally asserted that the treadmill portion of the experiment cost virtually nothing and that it was only a tiny piece of a much larger study of the effects of water quality on a fundamental link in the entire global food chain. Their vehement "denials" only cemented the public's perception that scientists had once again been caught red-handed wasting taxpayer money.

Appeals to aesthetics, awe at the immensity of the universe, and the sheer wonder of discovery continue to fall flat. Some have tried comparing numbers instead, showing that the money shelled out by the federal government to fund science research pales in comparison with other expenditures. Astronomer and de facto science ambassador Neil deGrasse Tyson famously and passionately argued this point on *Real Time with Bill Maher* a month after Congress had announced its cancellation of the James Webb Space Telescope. "Do you realize that the $850 billion TARP bailout [legislation passed in October 2008 to strengthen the faltering U.S. economy] . . . is greater than the entire 50-year running budget of NASA?" he asked animatedly. "It's not that you don't have enough money. It's that the distribution of money that you're spending is warped in some way that you are removing the only thing that gives people something to dream about tomorrow!"

And yet the discontent grows. According to Alan Leshner of the American Association for the Advancement of Science, a "substantial tension between science and the rest of society" has developed over the past few decades. At this writing, the negative sentiments have escalated from the level of grumbling comments about individual studies to the very process of national science funding itself. In April 2013, the United States House Committee on Science, Space and Technology modestly proposed the idea that the National Science Foundation require funded projects to have at least some promise of immediate benefit. Specifically, Chairman Lamar Smith stated: "It is the responsibility of the professionals at the NSF to exercise their best judgment and ensure that only proposals that benefit the taxpayer get funded."

Within weeks, the Canadian National Research Council followed suit. Their president, John MacDougal, stated unambiguously that "scientific discovery is not valuable unless it has commercial value."

Scientists are left scratching their heads at statements like this, which are often spoken as though every other human endeavor has an immediate benefit and that science alone is an impractical waste of time and money. But what about the apparent impracticality of the multibillion-dollar industries of music, movies, and literature? Most people agree that life would be far less rich, even less human, without the arts.

"But those are different," is the response. "I can choose whether to pay for music or movies or books. I'm forced to fund NASA."

Technically, this isn't completely true. Taxpayers fund the National Endowment for the Arts to the tune of $150 million each year. Although it's a lottery sum most of us dream of winning, that's virtually nothing compared with the over $3.5 trillion the United States spent in 2012, a number that is projected to increase every year. That's $3.5 *million* million. This figure is fully $1.1 trillion more than the United States pulled in that same year, which is why there is such a great cry to reduce spending. Any spending. The National Science Foundation makes a fairly easy target with an annual budget of about $7 billion, which is an enormous-sounding number, far greater than the NEA's funding. But even that enormous-sounding number amounts to only 0.2 percent of the nation's expenditures. With a budget of nearly $18 billion, NASA is funded by a mere half a penny on every tax dollar, and less than a third of that is dedicated to its science arm. Overall, total nondefense federal funding to science agencies amounts to approximately $70 billion per year, and less than half of that is devoted to basic research (the rest funds applied research, facilities, equipment, and development). That means that a mere 1 percent of overall government spending funds the pure scientific pursuit and creation of new knowledge. Even if we cut all science funding (which nearly everyone agrees would be an unwise move), the United States would still be looking at about a trillion-dollar annual deficit.

So let's look at another industry that is heavily supported by tax-payers: the National Football League. Between 2000 and 2009, tax-payers in 14 cities collectively forked over more than $4 billion to have stadiums built so that a subset of those same people could then pay an additional $75 per ticket to see a few dozen elite, heavily clad multimillionaire athletes wrestle each other back and forth across a specially tended field for three hours.

To put that figure into perspective, the total population of the 14 cities receiving new stadiums over that decade was about 25 million. This translates to a per capita tax burden of $16 per year to fund the construction of these stadiums. In an interesting coincidence, this figure—$16 per year per person—is almost exactly what it takes to fund NASA's entire science program. And yet, while there are plenty of science enthusiasts who welcome a new stadium so that they can paint their faces on the weekends and scream their teams to victory (it's true; I've seen them), it seems a bit harder to find an average football fan who wants to foot the bill for understanding the role of dark matter in galaxy formation.

But why? In a world filled with impractical diversions, what is it in the announcement of the discovery of a remote planet or a new particle or the surprising behavior of some undersea oddity that causes so many to attack with such venom? Isn't the "wow" factor of discovering something new about our universe just as practically important as the chance to root for the home team or argue with the referee over a bad call?

The truth, as revealed so frankly by the Canadian National Research Council Director, is that for many it all comes down to the bottom line. It becomes not so much a question of "What good is it?" as it is "What good is it *to me*?" Taxpayers in the regions that scored new football stadiums will tell you that they had their eyes on tomorrow: future revenues, new job prospects, and revitalizing the cities. What's more, they had personally seen the types of jobs that were needed to create such a structure. They knew that the stadium would be populated for decades with employees selling tickets and hot dogs, seeing people to their seats, or just running the day-to-day operations. It was more than simply lining the pockets of the multimillionaire owners and players. It was personal, and it affected the lives of people around them in very obvious ways.

But shrimp on a treadmill? Or some planet discovered around an impossibly distant star? Or the particle that took a $10 billion collider and an army of physicists to discover? What can these seemingly random things ever do to improve the life of the average person? What can any of this funding for curiosity-driven research do to help with the practical problems faced by billions of people around the world on a constant basis?

These questions are not new, nor are the answers. Kepler's response appealed to the innate human drive to understand the universe. To paraphrase pop icon Lady Gaga, humans were born this way, and fighting this natural curiosity is pointless. Michael Faraday, whose nineteenth-century discoveries about electromagnetic induction now help start your car, was a touch more practical and gave "an answer to those who are in the habit of saying to every new fact, 'What is its use?' Dr. [Ben] Franklin says to such, 'What is the use of an infant?'" Left unsaid was that someday the baby will grow up and contribute to society. Still, some take a more cynical approach. Tiring of the complaints that his work was "useless," J. J. Thomson, whose discovery of the electron earned him the 1906 Nobel Prize in Physics, would often toast, "To the electron! May it never be of any use to anybody!"

Beyond a laundry list of staggering scientific accomplishments, Thomson left us with a profound lesson: scientists are truly awful at gauging the ultimate practicality of their own discoveries. The properties of this "useless" electron are being exploited all around you, every minute of every day (even during power outages).

The idea that federal dollars are being wasted by scientists was fully politicized in the 1970s and 1980s by Senator William Proxmire. He created the Golden Fleece Awards, which were bestowed on publicly funded endeavors that appeared from his seat to be a complete waste of time and money. One of the first projects in his crosshairs was NASA's Search for Extraterrestrial Intelligence, or SETI, a $2 million per year project that doesn't look quite so far-fetched now that we are discovering new planets every week. Another was a $250,000 grant from the 1950s to study "The Sexual Behavior of the Screw-worm Fly."

As far as Proxmire could tell, this USDA-funded "science" was nothing more than creepy voyeurism masquerading as scholarship. Unfortunately, he judged this and many other studies by their titles, a habit picked up by many "journalists" who found paying jobs pointing out seemingly wasteful government spending. As it turns out, though, the screw-worm flies lay their eggs in open wounds of livestock. The maggots literally consume the flesh of the host animal for a week, resulting in gruesome infections and, very often, agonizing death. Understanding the reproductive cycle of these nightmarish creatures was crucial to ultimately controlling them. And controlling them saved literally billions of taxpayer dollars while reducing the suffering of countless animals affected. Not a bad return for a quarter of a million dollars spent in the 1950s. Senator Proxmire would later apologize for having disparaged this study, and by 1988, the Golden Fleece Awards had run their course.

But the seeds of public mistrust in science had been firmly planted, and they were vigorously cultivated by an economic downturn.

All is not lost, though. Recently U.S. Congressman Jim Cooper, with the enthusiastic support of virtually every major science body in the nation, instituted an effort to reverse that trend in perception. Instead of Golden Fleece, we now have Golden Geese. Or, rather, the Golden Goose Award, an actual honor awarded not to the apparent money wasters but to the surprising game changers. They look beyond the titles and questionable press releases and get to the heart of the science being done. For instance, it was easy to ridicule the scientists who made glow-in-the-dark cats. They were a brief Internet sensation around 2007 that drew the ire of virtually every commenter, who claimed that (1) it was cruel and (2), even if it wasn't

cruel, the link between glowing kitties and anything remotely ben-
eficial was tenuous at best (many commenters claimed that scientists
were only making up stories about how it might help cure AIDS
just to shut up the public). Behind these fluorescing felines, though,
was an exploration into the mechanism through which some jellyfish naturally fluoresce. This led to tests with genetic engineering in
mammals and has since resulted in an unprecedented ability to map
specific chemical activities in the human body. We now are seeing
major breakthroughs in medical fields never dreamt of when Osamu
Shimomura and Frank Johnson were first curiously gathering their
glowing jellyfish samples in the early 1960s. That initial curiosity,
and the decades of follow-up studies, merited a Golden Goose.

It seems, then, that the underlying problem is not so much that
curiosity-driven science is a waste of money but that people can't
readily see the return on their investment the way they can with a
new football stadium. This is partially the fault of scientists, who
have collectively dropped the ball in communicating the legacy of
science in an effective way. But it's also symptomatic of a society
that has so thoroughly compartmentalized education that we almost
never learn of the cross-discipline links that have been so vital to our
cultural existence. On top of that, we have also become an impatient
society, one that grows frustrated with microwave cook times and
that abandons a website if it takes more than a few seconds to load.
It took 50 years to transform the gee-whiz curiosity about glowing
sea creatures into useful therapies, an eternity to a culture that has
come to expect all things to be fixed in a four-year election cycle.

Unfortunately, we have begun applying that sort of short-term
planning to our own lives. A decade or two of hard work is a long
time to wait for an uncertain payoff ("What if I *don't* become a
successful musician?"), hence the almost cancerous spread of get-
famous-quick television shows that promise that we, too, can be
"discovered" without appearing to invest much time or effort. Why
can't life-changing breakthroughs be so easily made?

More important, though, some ask why our government—and,

by extension, the populace—has to pay for this stuff. If science is so lucrative, why don't private corporations foot the bill and make a handsome profit from curiosity?

For one, companies have surprisingly short lives. Most companies peter out within the first decade. (It is interesting to note, and not just a coincidence, that companies that got their starts in university research have substantially longer average life spans.) But even wildly successful companies that grow into multinational corporations have an average life span of only about 40 years. Sure, that's ten presidential elections, but it's not enough time to go from "Wow, neat jellyfish. I wonder why it glows?" to "We can now use this knowledge to treat genetic disorders." For another, companies have specific missions, and the people within them generally have even more specific job descriptions. It would go against that mission if the executives of, say, Frito-Lay suddenly decided to fund a research arm to explore undersea vents simply because that's what one of their employees wanted to do. Soon their shareholders would let them know what they thought of that move. Research that ate into this quarter's profits would be closely scrutinized, regardless of its alleged potential for a 50-year payoff. Rare is the company that has an entire arm devoted to unrelated and seemingly impractical research. (Rare, but not completely absent; Fry's Electronics, for instance, funds the American Institute of Mathematics purely to allow top mathematicians to perform their mental gymnastics.)

Governments, in contrast, are in the business of keeping society going from one generation to the next, with the idealistic dream that a great nation can thrive forever. In theory, all governments should have a long-haul mentality, acting as a good parent who doesn't allow her toddler to eat candy all day because she knows that will cause the child serious health problems as an adult. In practice, though, governments also have their own sorts of "shareholders" in the form of voters, and those voters often throw tantrums that end the careers of politicians. But, just as shareholders in a company should keep abreast of that company's activities, citizens of a nation have a responsibility to understand what activities their government is supporting. Instead of demanding that scientists be forced not to pursue seemingly pointless questions, people should ask themselves why the government feels obliged to fund pure science research in the first place.

And it's not just one government or two. Across the globe for virtually the entire history of science, the powers-that-be have supported scientists in a purely intellectual exploration of our world. This type of curiosity-driven exploration led to such bizarre findings as infrared light, discovered by a very surprised Sir William Herschel on the regal dime (or pound) of King George III in 1800. A mere curiosity at the time, infrared light is used regularly to save lives in the twenty-first century. With it, firefighters can "see" through the smoke to find people in burning buildings, and search and rescue teams can locate lost hikers on cold, impassable mountains. In 2013, a fugitive bombing suspect was found hiding under a boat tarp in a Boston suburb, thanks to the application of science funded by an eighteenth-century monarch.

It is almost certain that Herschel never included a statement of "broader societal impacts" in his proposal to toy with sunshine. In fact, until fairly recently, such language was absent from science proposals to national funding agencies. As Ronald Hoy, a professor of neurobiology and behavior at Cornell University, points out, "Back in the 1970s, the pure pursuit of basic questions in biology was a legitimate reason for funding, and we built nice research programs for several decades. But at the turn of the century, things changed and words like 'translational,' 'mission-oriented,' and 'deliverable' crept into biology."

Of all people, Dr. Hoy should know the value of curiosity. His discovery that a parasitic fly, whose sound sensory organs should not be large enough to discern direction, manages to locate its host cricket has great potential in the hearing aid industry. Helping those with hearing loss have better directional hearing was never on his radar, though. He simply is interested in the songs of bugs, an impractical pursuit if ever there were one.

If it's unreasonable to have expected Hoy to realize in the 1970s how his work would impact the hearing aid industry in the 2010s, then expecting Herschel to have had even the remotest grasp of where his work on light would ultimately lead would have been insane. Yet more than two centuries later, it has become practically mandatory for scientists to know what impact their work will have, in both the long run and the short run. Forget adding to human knowledge. What will its immediate payoff be? As J. J. Thomson's toast to the electron has taught us, scientists simply don't know. They *can't* know. In the words of renowned rocket scientist Wernher von Braun, "Basic research is what I am doing when I don't know what I am doing."

What history has shown us, though, is that our combined intellectual activity across disciplines, eras, and cultures is like an unfathomably big crossword puzzle. You can remain stumped for ages on 16 Down, but when you finally figure out 11 Across, a whole block of the puzzle, including 16 Down, suddenly becomes obvious. If the current trend continues, we will arbitrarily erase 11 Across because the citizens of our nation—our governmental "shareholders"—deem it unimportant. Unfortunately, without it, we could be missing a vital link to a large piece of the puzzle. If that piece happens to include your life, or the lives of your loved ones, it can make all the difference.

In the following pages, you will read some real-life dramas plucked from the headlines. Along the way, you'll learn about the work of just a few scientists whose goal was not to play a part in those headlines but to explore a question they simply thought was intriguing. And, although they often failed to answer the question, they never failed to have a dramatic impact on our lives.

PART I

✱ ✱ ✱ Finding Ourselves

IMAGINE. You are awaiting a heart transplant. It's agonizing enough to think about waiting for your own, but, when the intended recipient is your own child, the anguish is multiplied. You remain vigilant, ready. You don't stray too far from the hospital. You hope against hope. But you also try to live. Going to a concert seems to be the perfect celebration of life and humanity while also being a great escape from the waiting and wondering. And normally it would be, but during this celebration the doctors try to contact you. A matching organ has been found. There are only hours to get it to the recipient. Your child. When there is no answer at your home phone, they try your cell phone. No answer there, either. A considerate patron, you've turned off the ringer for the concert but thankfully not the entire unit. Undeterred, the doctors contact the police, who canvas the area trying to find you. One officer suggests tracking your phone's GPS signal, which pinpoints you in the concert venue. The concert pauses. An officer takes the stage, while the hushed audience wonders what is wrong. He makes an announcement that a donor heart has been found for your child, and suddenly everyone—including the musicians—is cheering for your child, who has a new lease on life. Amid the joy and chaos, nobody wonders or even much cares how it was possible to locate you so precisely. Finding you, however, hinged on the century-old mental musings about the nature of time and space and on generations of precision experiments devised to find out if those musings were right.

1   A BRIEF HISTORY OF TIMING

The modern Global Positioning System (GPS) is practically ubiquitous these days, so ubiquitous, in fact, that the thought of being without the technology to find your exact location is practically nonsensical. What is easy to forget, though, is that this ability to find yourself—whether you're in a concert hall, driving down an unfamiliar road, or sitting in your living room—is made possible only by an amazing triumph of modern technology: an armada of satellites orbiting Earth. And their ability to do their jobs came about because for the past century, humans have explored the precise interaction between time, space, and matter, an interaction so bizarre and so seemingly impractical that the question "What good is it?" is one of the first things people ask when they encounter Einstein's theory of relativity. Without these investigations into the fabric of the universe, however, those satellites would be practically useless.

At any given time and at any given location, at least six of these satellites have a direct line of sight to you. Contrary to popular belief, these satellites are not in geosynchronous orbits—"staying put" above a certain point on Earth as the planet rotates—but instead orbit at an altitude of 20,200 kilometers (12,500 miles). At this distance, each satellite naturally makes two full orbits of Earth each day. With six orbital planes and four satellites in each (there are actually 30 GPS satellites in orbit, but only 24 are part of our everyday navigation system), the result is a hypnotic and rhythmic choreography of technology, the animation for which certainly merits at least a cursory Google search.

While engaged in their geocentric dance, these satellites (known to those in the industry as "space vehicles") are constantly sending out signals with the exact time and their identity encoded within. It's this speed-of-light signal that the receiver in your car or phone or other GPS device picks up, but your receiver's clock isn't nearly as accurate as the atomic clocks aboard each satellite. Although most people typically aren't concerned with a watch that gradually gains or loses a second here and there, even imperceptibly small inaccuracies can create huge problems where the speed of light is concerned. For instance, a minuscule discrepancy of a thousandth of a second in your receiver's clock could create an uncertainty in your location of 300 kilometers (186 miles), the distance between New York City and Baltimore, Maryland.

Obviously, this sort of ambiguity is not going to be much use in pinpointing your location, but even more impractical—not to mention exquisitely expensive—is equipping your receiver with a similarly accurate clock. The solution is to have your GPS receiver pick up signals from at least three different satellites—typically four to seven—to get a good read on your position. With the combined time and position signals from multiple satellites, the accuracy of the receiver's clock is no longer critical, and it can easily compute your position to well within 100 meters.

Notice that the whole process hinges on precision timing, a feature that is easy to overlook. Most descriptions of the satellites will tell you that they are equipped with atomic clocks but tell you nothing about what an atomic clock is, why it's so precise, or what drove humans to try to create a clock that gains or loses only a second every hundred million years.

This obsession with timekeeping isn't anything new, though. Ancient schedules revolved around annual, seasonal, monthly, or daily rhythms, and innumerable examples of timekeeping structures and rock carvings from these early cultures still pepper our planet in famous places like Stonehenge in Wiltshire County, England, and in less famous places like the V-V Ranch Petroglyph site near Sedona, Arizona. Other cultures, like the Babylonians and Egyptians, subdivided not just the year but also the day into units by marking the path of the Sun with a gnomon or sundial. And as early as the third century BC, the rigors of a compartmentalized daily schedule had driven Roman dramatist Titus Maccius Plautus to write passionately against this practice:

> The gods confound the man who first found out
> How to distinguish hours! Confound him, too,
> Who in this place set up a sundial,
> To cut and hack my days so wretchedly
> Into small portions.

Little progress in the basic technology of time keeping occurred for centuries. Elaborate sundials replaced the crude monoliths of ancient times, and people came to realize that a sundial angled parallel to Earth's axis of rotation yielded a uniformly moving shadow and an opportunity to cut the days up into even slices. Weight-driven mechanical clocks, water clocks, and hourglasses were developed, but these devices suffered from occasionally enormous inconsistencies.

It wasn't until Galileo's observation of swaying suspended lamps in the cathedral of Pisa that an entirely new way of keeping time was born, one where every time interval was the same as the one before and one by which even small intervals of time could be measured. What Galileo noted about the lamps was that the time for them to swing completely through an arc did not depend on the width of the arc. A lamp swinging through a smaller arc took the same amount of time as one swinging through a larger arc. An interesting tidbit: Galileo used his own pulse rate as the timing device to determine this amazingly useful property of pendulums. He then explained the isochronic property of swinging weights to a physician friend who immediately employed it to measure the pulse rates of his patients, making it quite possibly the fastest "curiosity-driven discovery to practical application" story in history.

Despite its application to music (the metronome) and medicine, the ability to measure small time intervals with a great deal of accuracy was a boon mostly to astronomers, who spent the next few centuries pushing the limits of chronometer precision. By the 1920s, a pendulum clock precise to one part in 30 million (i.e., it gained or lost only a second per year) had been developed by William Shortt. This might be time-keeping overkill for most of the human race but was still not good enough for the challenges of science. This clock was so sensitive that its performance was affected by the subtle gravitational variations arising from tidal distortions as the Moon and Sun ever so slightly rearranged the distribution of Earth's mass.

The key to the precision of any clock is a reliable oscillator. A pendulum, for instance, could be designed to have an oscillation frequency of one hertz, or one full swing (to and fro) per second. If you assume that your pendulum is completely reliable, you simply gear your clock so that the minute hand moves one tick mark forward for every 60 pendulum swings. But what if the pendulum's frequency goes down for some reason? Then 60 swings will actually take more than a minute, but your clock will still measure only a minute. Unbeknownst to you, your clock will slowly fall behind. The problem with the pendulum clock—and any man-made timing device, for that matter—is that it inevitably suffers a gradual change in its properties. In other words, things eventually wear out.

So what doesn't wear out? Fully 50 years before the Shortt clock was developed, James Clerk Maxwell—a scientist whose entire career was spent doing research at publicly funded institutions in Scotland and England—realized that atoms themselves might hold the key to precision timing. Unlike pendulums, atoms of a given element are amazingly uniform and, as far as we know, they never change what they do. A hydrogen atom, for instance, typically consists of a single proton and a single electron. When energized, a hydrogen atom will emit very specific frequencies of light, frequencies that will be the same whether measured yesterday, today, or tomorrow in a lab in Sydney, in Seoul, or on Saturn. Moreover, the frequencies are extremely high—hundreds of trillions of oscillations per second for the visible emission lines of hydrogen—and in theory could yield a precision of better than a hundredth of a trillionth of a second. Maxwell favored sodium atoms for the task, declaring with just a hint of biting sarcasm directed at the eminent scientist Lord Kelvin:

> The most universal standard of length which we could assume would be the wavelength in vacuum of a particular kind of light . . . Such a standard would be independent of any changes in the dimensions of the earth, and should be adopted by those who expect their writings to be more permanent than that body.

Lord Kelvin, whose enduring legacy includes an entire temperature scale, realized that, despite the longevity of all his personal accumulated accomplishments, Maxwell was right. However, beyond just the practical problems of attempting to count such rapid oscillations, an overarching difficulty with using an atom as a basis for measuring time is that atoms are so easily perturbed. Just as a train whistle seems to change pitch as it approaches and recedes from you, the observed frequency of light emitted by an atom in motion will differ slightly from the observed frequency emitted by one at rest. The faster the atom, the bigger the Doppler shift, named after Christian Doppler. (A personification of scientific networking, he also happened to be one of Gregor Mendel's physics teachers at the University of Vienna and we will meet him again later in the book). Atoms are pretty much always in motion, though, and room-temperature hydrogen gas has particles moving up to thousands of kilometers per hour in various directions. This motion creates a wide range of measured frequencies around the "rest" one. Other bizarre and less intuitively grasped factors internal to the atom also create a small spread of frequencies around an atom's particular resonance frequencies. All of this makes it virtually impossible to achieve the mind-boggling accuracy of one second for every three million years.

But why would minds like Kelvin, Maxwell, and Doppler even bother thinking about that kind of clock?

2   GOING WITH THE FLOW

Matter. Time. Space. The question of how these are related, if indeed they are, has been asked since ancient times, but rarely has any answer been based on anything other than a "gut" instinct. Aristotle posited that matter and time and space were all interrelated through the phenomenon of motion. Without matter to do the moving, and space for that matter to be moving through, the concept of time was pointless. And for all of his amazing achievements in physics, Isaac Newton just "knew" that time and space were absolutes. Or, as he more eloquently put it in his 1687 opus on all of classical mechanics, *Philosophiae Naturalis Principia Mathematica* (those in the know call it simply the *Principia*): "As the order of the parts of time is immutable, so also is the order of the parts of space."

For two centuries, everyone was content to trust Newton on this one, since his theory of gravitation and laws of motion had stood up to just about every test physicists threw at them. With the stable footing of Newton, physics was ready to enter new territory: discovering the absolute speed of Earth through that immutable space. People understood the speed of Earth as it revolved around the Sun and also realized that Earth rotated on its axis, giving everything on Earth an imperceptible, but very rapid, motion. What Newton's laws could do very well was describe how objects moved relative to each other, but where they failed was in showing us that universal grid through which we must be moving.

By the 1880s, though, science seemed up to the challenge. At that time, it was well established that light behaved like a wave. It could be bent and reflected and even interfere with itself, producing "high" points and "low" points (intense spots and shadowy spots) where light waves added up or canceled out. Since every known wave phenomenon had to be riding on some medium, then light, too, must be traveling on some invisible carrier. The properties of that medium could be sneakily divined not by observing it but instead by observing the behavior of the thing that travels through it: light itself.

For one thing, light moves very quickly. By the 1870s, Leon Foucault had determined its speed to be nearly 300,000 kilometers per second with an amazingly clever setup that involved bouncing light waves off a rotating mirror. Physicists had also figured out that visible wavelengths were somewhere around a millionth of a meter, meaning that the frequency of the wave had to be hundreds of trillions of cycles per second, or several hundred trillion hertz. Whatever light rode on, it could support a very fast wave and it could oscillate very quickly. What kind of medium could possibly do that?

A guitar string provides some clue. If you pluck a guitar string, it makes a particular note. If you twist the tuning peg and increase the tension, the same length string will make a higher-pitched note. The only difference is the tension, and yet this makes the frequency increase. This observation implies that the waves themselves must be moving faster along the tenser string. Extending this finding to light waves forced physicists to the uncomfortable conclusion that whatever its medium was, it possessed a higher tension than any known substance.

Such a rigid medium made no sense, however, because clearly everyday objects passed through what was then called the luminiferous, or light-bearing, ether without any trouble. So the medium was rigid, but completely lacking in any viscosity or drag, at least for objects moving at much slower speeds. Furthermore, it filled every nook and cranny of space, as evidenced by the fact that we could see the light from distant objects. So this mysterious medium had to be fluid as well. On top of it all, the ether was totally transparent. It was beginning to look tragically like this ether was the substance of the emperor's new clothes, but it was almost crazier to believe that light waves could somehow travel without an appropriate medium to travel through.

In 1887, Albert Michelson and Edward Morley set out to determine, if nothing else, the speed of Earth through this ether. If it were truly the medium that light waves danced across, then Earth had to be plowing through ether all the time. Earth's orbital speed is comparatively tiny—only about 30 kilometers (18 miles) per second, as opposed to light's 300,000 kilometers (186,000 miles) per second. But even this tiny speed should have measurable effects, assuming that light always moved at the same speed through its ether, an assumption supported by James Clerk Maxwell's famous equations from the 1860s. In these four equations, Maxwell codified virtually all of electricity and magnetism. Light, as it turned out, was an electromagnetic wave, and its constant speed was one of the natural consequences of Maxwell's equations. But speed relative to what? It must be the ether, everyone declared confidently.

Since it was highly unlikely that the ether and Earth were traveling in tandem, our motion through the medium should be discernible. Michelson and Morley figured that a light beam fired "upstream" into the direction of Earth's motion through the ether should move more slowly than the advertised speed of light, just as someone swimming against the current of a river will have a lower speed relative to the banks while swimming upstream. Michelson even explained the basic idea of his experiment to children, describing how a swimmer that swam first upstream and then downstream should take slightly longer than one who swam to the opposite bank and back, even though their relative distances were the same. The conclusion was simple enough even for a child to grasp, and there was no reason for anyone to think that beams of light shot into the ether should act any differently from swimmers in a river. A beam of light fired "upstream" (in the direction Earth was moving) for a certain distance, reflected, and then sent "downstream" back to its starting point should take slightly longer than one fired "across the river."

Since the speeds were too great simply to clock the difference in arrival times of the light beams, Michelson and Morley had to make use of light's interfering properties. As with any wave, when the crest of one light wave meets the trough of another, the light will cancel out. But if two crests meet, there will be a bright spot. What this means is that if you send out a single light beam and use a partially reflective surface to split it into two perpendicular ones, the two new beams will start out in phase; crest will match crest and trough will match trough. If one beam is then sent upstream through the ether, its return should be slightly delayed. With the proper arrangement and fine-tuning of mirrors and detectors, Michelson and Morley figured it should be possible to change the light beams' interference pattern.

Drawing such an experiment out on paper was one thing. The practical aspects were another beast entirely. Michelson had tried it alone in 1881 with inconclusive results and imprecise tools. He wrestled with the problem of light endlessly. His daughter, Dorothy Michelson Livingston, recalled, "When someone would ask why he spent his whole life on the behavior of light, I saw his glowing smile as he replied, 'Because it's so much fun.'"

As fun as it was, Michelson suffered a nervous breakdown in 1885. Soon after, he enlisted the help of chemist friend Morley to create something so precise, so immune to outside disturbances, that Earth's motion through the stationary ether should become obvious. The final setup involved a gargantuan rotating rock platform floating in a trough of liquid mercury, an apparatus made possible because of a modest grant from the National Academy of Sciences, then only two decades old. Michelson himself would later become president of this body, which was founded to provide scientific advisors to the government and, in return, to provide funding to scientists so they could explore topics they found intriguing. For Michelson and Morley, that meant exploring the ether, and they painstakingly sought any sign that one light beam arrived a millionth of a millionth of a second faster than the other.

And still, they failed to find the elusive substance.

What they did discover, though, was that no matter how they

set things up, whether the light beam was sent against the direction of Earth's motion or perpendicular to Earth's motion, whether they tried the experiment in summer or in winter, at night or in the daytime, the result was the same. The light beams always came back in phase with each other. Several reasons were suggested for this unexpected null result, including the possibility that Earth dragged the ether with it somehow, or that the ether itself compressed the apparatus in the direction of motion, making light's path just the right amount shorter to go there and back in the same amount of time. For the next decade, a flurry of famous physicists would frantically struggle to explain how light could always appear to travel at the same speed, regardless of the speed of its source.

No one would solve the puzzle as elegantly and as counterintuitively as Albert Einstein.

3   FROM *PRINCIPIA* TO PRINCIPE

The answer was actually quite simple, requiring only that physicists abandon much of what they always held as truths. Since an object's speed is nothing more than the relationship between the distance covered and the time it takes to cover that distance, the only way to have the speed of light universally constant was for distance (i.e., space) and time to be dependent on the observer and, more disturbingly, on each other. Newton's absolute space and time with their rigid, equally spaced gridlines were an illusion, and everyone's strange experimental results could be easily explained if only we could persuade ourselves to stop thinking in ancient absolutes.

It isn't easy to do.

Einstein's 1905 paper set down what would later be called his special theory of relativity, special because it was applicable only to objects moving at constant speeds in straight lines. Einstein knew that the scope of this paper was too limited, though, so he began to tackle the problem of accelerated motion before the ink was even dry on his first paper. Then, in 1907, while working in the patent office of Bern, Switzerland, Einstein had what he would later describe as "the happiest thought of my life." It would become his equivalence principle, and it would help link all motions and masses to a single enormous idea: the general theory of relativity.

The most obvious accelerated motion that every earthling encounters is that of a falling object. For more than two centuries, physicists had explained falling with Newton's law of universal gravitation. This law had done an amazing job of describing quite accurately the behavior of everything from falling cannonballs to the orbits of the planets. Gravity was a force, like the attraction of two opposite magnetic poles or the attraction of two opposite electrical charges. But somehow, unlike the case of magnets and charges, different masses had resulting motions that were always the same. A marble lands at the same time as a cannonball dropped from the same height, always accelerating at the same rate, indicated by the letter g, to the surface of Earth.

Once again, Einstein would think differently about the problem, looking at it not from the grounded reference of an earthbound spectator but from the viewpoint of the falling object. His happiest thought had actually been the stuff of many people's phobias: "For an observer falling freely from the roof of a house, the gravitational field does not exist." If you have ever found yourself strapped into a free-fall ride at a theme park, lingering at the top as you wait for the inevitable drop, Einstein's thought experiment might have occurred to you. One of the more popular ways to distract yourself from the terror of plummeting to the ground below is to rest a penny on your knee and concentrate on it. What you witness looks somehow unnatural—the apparent cessation of gravity.

It is such a simple observation, really. While you and the penny fall, the penny seems to be weightless, Abraham Lincoln floating like a tiny astronaut above your knee. This is no different from the way the penny would behave if you were strapped in an identical chair in the depths of interstellar space, far from any planets or stars that could weigh you down. By extension, the sensation of earthly gravity should be no different from being in the depths of interstellar space inside a rocket that is going 35 kilometers (20 miles) per hour faster with every ticking second. In other words, in a rocket accelerating at exactly one g. Gravity, Einstein reasoned, is not some mysterious force that reaches out across space and pulls us to the ground. Instead, it is the natural consequence of the interaction of matter, space, and time.

Even as early as 1907, Einstein realized that this universe of interdependence would have some interesting observable side effects, but unfortunately he lacked the mathematical skills to formalize them.

The idea to weave space and time into a single fabric came instead from Einstein's former math professor, Hermann Minkowski, a man who had called Einstein a "lazy dog" who "never bothered about mathematics." Minkowski could see mathematically what his lazy dog student did not, and in September 1908, just months before his death, Minkowski announced dramatically to the 80th Assembly of German Natural Scientists and Physicians: "Henceforth space by itself, and time by itself, are doomed to fade away into mere shadows, and only a kind of union of the two will preserve an independent reality."

Einstein himself initially thought that Minkowski was simply enshrouding his beautiful physical insights within ugly mathematical formalisms. There seemed to be at least some mutual envy: Einstein had a stunning intuition about the universe, and Minkowski had a gift for seeing the mathematical order in that universe. By the time he started grappling with accelerating motions, Einstein was forced to warm up to the strange four-dimensional space-time and the esoteric mathematics required to describe it.

Rather than a three-dimensional space filled with Newton's rigid universal gridlines and a separate arrow of time, Einstein's universe was more fluid, bendable even, and all of the space and time gridlines were interwoven. What we think of as an acceleration downward to the ground is actually the result of warped time. The time "gridlines" aren't uniformly positioned around Earth from our point of view, but to a falling object, they are. An object blithely moving from one time gridline to the next at what it perceives be a uniform rate will appear from our trapped three-dimensional perspective to have an accelerated motion. By 1911, Einstein had finally worked out some of the mathematical issues, publishing in June of that year the first hints that our experience was not as trapped as it appeared.

But why bother? Was it just a matter of having an overarching set of equations so that a cadre of elite university-funded physicists could finally describe motion more precisely than anyone would ever need to? Or was it a constant game of one-upmanship by mental gymnasts? What difference could it ever make to the average person that falling objects aren't really being pulled down by a force of gravity? Would it have helped any of the 148 girls and women who, in March 1911, leapt to their deaths from eighth, ninth, and tenth floor windows of the Triangle Shirtwaist Factory as it burned? In a just world, how could a man who devoted his time and energy to something so unreal and impractical be rewarded with fame, while those trying to make an honest living were rewarded with nightmarish death? Who cares if gravity is a force or a warp in space-time? It certainly didn't matter to those who experienced its unrelenting effects for 30 meters, straight down to the hard ground below.

Even for those who don't face such harsh trials, it's much easier to think of motions and forces and dispense with this strange notion of space-time. Most physics classes still ignore it entirely and teach Newton's mechanics virtually unaltered from his 1687 work *Principia*. Time is one thing in those classes. Space, another entirely. In fact, space is even further shredded into its three-dimensional components, each independent of the others. After all, we can't see space-time scrunched up in one spot and stretched out in another, so it makes no sense to complicate matters by including those subtle effects.

But Einstein's 1911 paper—a precursor to the full general theory of relativity that he published in 1915—made predictions about how warped space-time would affect things we can observe. For instance, just as glass changes the direction of a beam of light, sometimes resulting in multiple images of the same background object, a massive object should cause distortions in space-time that will make distant objects appear out of place. A light beam that passes closer to the massive object would be bent more than one passing farther from the object, and the most measurable effects would be caused by objects with highly concentrated mass. Inconveniently, the most massive object in our immediate neighborhood also has the distinction of being the brightest. Einstein showed that the Sun, although glaringly bright, has a predictable effect on the apparent arrangement of background stars—if only you could somehow shut it off long enough to see those stars.

As luck would have it, nature occasionally arranges for just such a solar blackout. A total solar eclipse, where the moon passes in front of the Sun, could in principle provide hard evidence for the theory of relativity. But an eclipse is not easy to come by. Earth experiences about one total eclipse per year on average, and even that can't be seen by everyone. Totality—where the Moon's disk completely blocks the Sun's—is only ever visible for a few minutes along a narrow band about 30 kilometers wide. After Einstein's initial prediction (and a subsequent factor of two correction; even Einstein had trouble keeping the idea of rigid gridlines from interfering with his computations!), he formalized his full theory in his famous 1915 paper. By this time, he had long since left his job as a Swiss patent clerk and had taken on the role of professor at Humboldt University in Berlin, a research institution where he professed little (his teaching duties were minimal) and thought lots, all with the financial backing of German society.

What his general theory of relativity needed was experimental verification in the form of a total eclipse. There were a few good opportunities from a viewing standpoint, but the start of World War I in 1914 put the 1916 and 1918 eclipse expeditions on hold. As it turns out, neither of these eclipses presented the ideal arrangement between the Sun, Earth, and background stars, so it's just as well that the expeditions didn't occur. To accurately and convincingly measure the bending of light, astronomers would need the Sun to pass in front of as many bright stars as possible. The perfect opportunity arose in 1919, when the Sun appeared to pass right in front of the Hyades star cluster, part of the constellation Taurus. With dozens of easily visible, well-mapped stars whose positions were known with incredible accuracy, the Hyades were the perfect laboratory to test Einstein's theory.

Unfortunately, this eclipse would not be visible from any part of Europe or North America where most of the interested scientists lived. In fact, the path of totality stretched from just west of Madagascar, across the uninhabited vastness of the Atlantic Ocean, all the way to the central western coast of South America. For six minutes—

an eternity for eclipse totality—mankind would get the opportunity to measure the positions of the stars of the Hyades cluster, if only the right people could make their way to a good viewing spot.

British Astronomer Royal Sir Frank Dyson and astronomer Arthur Eddington, who was desperately hoping to avoid military service in World War I, organized two expeditions to record the event. Eddington himself was required by the British government to head one of the expeditions in return for avoiding the draft. Fortunately, Eddington had already attempted such a measurement in 1912 (unsuccessfully; but since Einstein had not yet made the factor of two correction to his 1911 mathematical error, it was a good thing that Eddington had failed), and as a result, Eddington was deemed the most qualified individual for the job even before his request to be excused from military duty.

Finally, at 11 a.m. on the eleventh day of November 1918, World War I officially ended, and preparations for the expedition could be made in earnest. The plan was for Eddington to lead one expedition to the remote island of Principe off the coast of West Africa, while another astronomer, Andrew Crommelin, traveled with his team halfway around the world to Sobral, Brazil, for those six precious minutes.

The teams knew what was at stake. The Sun's gravity would do one of three measurable things to the light of the Hyades stars: nothing (which would have puzzled even staunch Newtonians), bend the light a small amount in accordance with Newton's laws, or bend the light by twice that amount, confirming Einstein's (corrected) bewildering theory of a malleable universal infrastructure. The shifts they were expecting were imperceptibly small. Light grazing the Sun's surface would appear to be deflected by less than a thousandth of a degree. This is about the difference between two adjacent squares on a standard chessboard as seen from a few kilometers away. And, as if that weren't problematic enough, there would be an enormous spotlight in the way.

On May 29, 1919, Eddington's team awoke to thunderstorms, certainly not the most auspicious way to start a day of precise observations of the Sun and faint background stars. The Sobral cohort, meanwhile, had aligned and adjusted their instruments the previous night. But in their eagerness to have everything just right, they failed to account for the expansion of the instrument in the daytime heat. By eclipse time, Crommelin's team realized their mistake but could not adjust in time. A last-minute backup telescope was used to capture some fair-quality photos, but all the exposures from the larger, specially designed instrument were omitted from the final analysis. Eddington's luck was only marginally better. He wound up shooting the eclipse through breaks in the clouds, enough visibility to see the Sun, and, he hoped, just enough to see the stars of the Hyades.

Eddington pored over his photographic plates while still on Principe. On June 3, he wrote in his notebook that his measurements had verified Einstein's relativity, a surprisingly bold statement considering his data. He returned to England apparently victorious, but it took months before all the photographs from both expeditions were fully analyzed. Meanwhile, he summed up his adventures poetically and presented at a Royal Astronomical Society dinner a version of Omar Khayyam's Rubaiyat that he wrote on the topic:

Oh leave the Wise our measures to collate
One thing at least is certain, light has weight
One thing is certain and the rest debate
Light rays, when near the Sun, do not go straight.

Einstein himself only found out the expedition's results via telegram in late September, a few weeks before the official presentation. As soon as he learned of it, he wrote not a poem, not a paper, but instead a brief, jubilant postcard to his mother.

In 1919 it was joked that only three people on Earth understood relativity, and Einstein and Eddington were two of them (arguments about number three were known to deflate many egos). Despite some troubling uncertainties in the results from Principe and Sobral, the astronomical community overwhelmingly agreed that the eclipse observations had verified Einstein's theory. J. J. Thomson, the discoverer of the electron who hoped his particle would never become useful, declared, "This is the most important result obtained in connection with gravitation since Newton's day."

At a time when nations were somberly picking up the pieces of a global war, news of a scientific breakthrough was like a beacon of hope. Life here might be a mess, but the universe is letting us in on its secrets, and, despite our problems, we are a species clever enough to decode those secrets. The eclipse findings were not shuffled into erudite journals but trumpeted around the globe. Einstein became an instant celebrity.

Eddington himself gave a number of lucid lectures on the subject, often to packed halls. One of these talks caught the attention of Cambridge undergraduate Cecilia Payne, who decided to pursue a career in astronomy after hearing it. Her decision, stemming from Einstein's musings over motion, would soon overhaul our understanding of stars. Payne was the first to understand the chemical makeup of the Sun, a breakthrough that, nearly a century later, is paving the way to a promising cancer treatment (see part V).

But that's not where this story leads.

4   THE ATTRACTION OF TIME

Although powerful in its own right, Eddington's result did not verify that time was wrapped up in this strange universal web of space-time. With even small portable telescopes, we could magnify the chessboard to resolve individual squares, even from kilometers away. But how could we ever hope to "magnify" time enough to measure how mass and motion affect it?

In Einstein's special relativity, speed itself could change the pace of time. With the requirement that everyone must measure the same speed of light, this meant that there would necessarily have to be some disagreement on distances and times. Fast-moving objects would appear shorter. Fast-moving clocks would appear to be ticking more slowly. How much shorter and how much more slowly was what Einstein figured out in his 1905 paper. For most objects, the changes would be invisible. An everyday object like a car cruising along the highway at 100 kilometers (60 miles) per hour is going about one ten-millionth the speed of light. A stationary hitchhiker watching a car pass would see something that was 99.9999999999995 percent as long as an identical car that stopped to pick up the hitchhiker. For your average car, that means about 25 femtometers shorter, or about 25 proton diameters, give or take. Imperceptible, in other words. Moreover, if both the cars had their turn signals on, the moving one's signal would appear to blink 0.00000000000049 percent slower than the stopped one's.

When things begin moving faster, though, the changes become more obvious. When the car gets to a tenth the speed of light—18,600 miles per second or 30,000 kilometers per second—its blinker flashes half a percent slower than the stationary one, and the car appears half a percent smaller. Now instead of a few proton-widths difference, the moving one will appear a couple centimeters shorter than the stopped one. A blinker flashing once a second will now seem to lose a flash every 200 seconds. This is something that can be measured fairly easily. Unfortunately, finding something that moves that fast is not quite so easy.

To discern these relativistic effects, physicists needed to find a clever way to clock something without actually measuring the "flashes." Those clocks could be found in the atoms themselves, as Maxwell suggested and generations of physicists had attempted to exploit. The hydrogen atom was known to emit a particular wavelength of light—486.1 nanometers, to be exact—and under the right conditions could certainly be made to go faster than a car. If you have an old tube television set or computer monitor, you've got the beginnings of the first instrument to test special relativity. Known as a cathode ray tube, these giant glass tubes are now amazingly fine-tuned instruments that could only have been dreamt of a century ago. When you turn them on, the voltage accelerates electrons that then smash into the front of your computer or television screen, causing the material there to fluoresce. With the right materials and tweaking of magnetic fields, an entire movie takes place on this screen.

Electrons aren't the only things affected by this voltage, though. Positively charged particles will be accelerated in the opposite direction as something called canal rays. A single proton—the basis of a hydrogen atom—can get up to speeds of half a percent the speed of light, give or take. At these speeds, relativistic effects start becoming measurable but only with a clever enough experiment. As formalized by Doppler, the motion of an object affects the wavelengths (frequencies) we observe. If the object is moving away from us, the wavelengths appear longer. If it is moving toward us, the wavelengths appear shorter. In a normal, non-relativistic world, those away versus toward wavelength shifts are symmetric around the "rest" wavelength. It doesn't matter which direction the object is moving. All that matters is the speed, and at the instant, say, an ambulance overtakes you, you hear the same note that you'd hear if it were simply sitting next to you. But in the bizarre world of relativity, the symmetry is gone, and the "rest" wavelength appears to be closer to that of the approaching object. Conveniently, all you have to do is look for the predicted asymmetry, rather than for some minuscule difference in frequency. Looking for that broken symmetry is precisely what Herbert Ives and G. R. Stilwell did in 1938 and again in 1941.

But this was only a small test of special relativity, and it didn't really measure the stretching of time so much as imply its existence. Einstein's general theory of relativity predicted entirely different effects related to gravity. Specifically, being closer to a massive object would mean that you were in a region of space-time where the universal gridlines were stretched out more. This would mean that a clock on the surface of Earth should keep time at a slower pace than one far from Earth, where the time gridlines were closer together. How much slower? According to general relativity, a year atop Mount Everest—all 8.8 kilometers (29,029 feet) of it—would be about 30 millionths of a second shorter than a year at sea level. To measure this tiny effect, you would need a clock that was more precise than anyone had even a generation after Eddington's expedition. And you'd have to get to the summit without breaking it. In this case, predicting such an effect was child's play compared with measuring it.

Naturally physicists did what physicists do: they kept atoms in mind, knowing that in these lay the perfect standard timekeepers. The wavelengths they had been using, though, were far too short to be of much use. Visible light waves are only a few hundred billionths of a meter long and have frequencies of hundreds of trillions of cycles per second. Nobody knew of longer, lower-frequency waves until the 1920s, when it was discovered that a very subtle change in the arrangement of an atom could give rise to microwaves. Microwaves, by contrast, have wavelengths that humans can appreciate, and even draw, ranging from about a millimeter in size to a meter. Their correspondingly low frequencies in the billions or millions of hertz are more easily manipulated, and it was in this region of the electromagnetic spectrum that the atomic clock got its first real foothold.

Israel Isaac Rabi, known to his friends as Izzy, led the first team to construct the precursor to an atomic clock. This device was made possible only because of a curious discovery about the magnetic properties of atoms and molecules. Just as current running through a coiled wire can create a magnetic field (a property capitalized on in your car engine's starter solenoid), the "spin" of charged subatomic particles can create a minute magnetic field.

A proton and an electron occupying an atom of hydrogen even have magnetic fields that interact with each other in very subtle ways. If the electron's spin is such that its magnetic field is aligned with the proton's, it's a bit like having two nearby refrigerator magnets with their like poles pointing the same direction. If left to their own devices, those two refrigerator magnets will spontaneously rearrange themselves so that the north of one is joined with the south of the other. On an unimaginably tiny scale, this is also what happens inside an atom. And, just as you can hear some of the energy that is released when the two refrigerator magnets clack together, physicists can see the energy released when the electron flips its spin. When it flips, it releases a very small amount of energy in the form a very specific frequency light.

What Izzy Rabi and his team did was to create a beam of molecular hydrogen (two hydrogen atoms joined together) and run that beam through a series of rapidly changing magnetic fields. These would essentially sift out the molecules whose electrons were in the "wrong" spin state, and only the ones that had achieved the desired magnetic orientations would be guided into a detector. By fine-tuning the behavior of the magnetic fields, Rabi could increase or decrease the number of molecules that ultimately hit the detector. Only by tuning to the right magnetic combination could he maximize the number of hits, and this magnetic combination revealed a wealth of information about the energy—and hence the frequency of light—involved in particular electron spin flips. These days, the entire field of magnetic resonance imaging hinges on the discoveries Rabi and others made on the taxpayer's dime during the 1930s. That's MRI in biomedical lingo, and it's a tool that discovered your teenage daughter's concussion after her opponent ran headlong into her during that soccer game last week.

But that's not where this story leads, either.

5   FINE-TUNING OUR CLOCKS

Rabi and his colleagues at Columbia University in New York later realized that the same physical underpinnings could exploit the now well-known magnetic properties and help create a frequency standard: an atomic clock. In 1945, Rabi gave a lecture to that effect, and the *New York Times* picked up the story and ran with it, boldly declaring "Cosmic Pendulum for Clock Planned—Radio Frequencies in Hearts of Atoms Would Be Used in Most Accurate of Timepieces." Had this article run on a current-day news website, the comment section would no doubt have been filled with questions about why our tax dollars should go to making a timepiece accurate to one part in 100 million when there was yet another world war to clean up after.

Rabi himself would not develop this particular chronometer. A number of groups wrestled with various substances and experimental setups, some with more success than others. In 1949, Harold Lyons successfully used natural electron transitions within the ammonia molecule to create a workable atomic clock. The apparatus itself looked like a large black box with dials, meters, and oscilloscopes—visually uninformative to the average person—so to make it look more clock-like for PR purposes, they added a normal electric analog clock at the top that served no actual function. Complications arising from the motions of the particles themselves made this particular clock no better than quartz crystal clocks of the era, but the idea worked. And every inventor knows that once you have a working prototype, removing the rough edges is comparatively easy.

(Speaking of working ideas, the quartz crystal technology has been going strong for nearly a century. These crystals, which oscillate at very specific frequencies when a small voltage is applied, are happily keeping time for you in cell phones, computers, watches, clocks . . . you name it. Their problem, as with any mechanical timing device, is that their properties change gradually as they grow older.)

Norman Ramsey at Harvard took up the atomic clock mantle next and added an additional level of filtering that did away with the Doppler effects. Then the choice to use the element cesium boosted accuracy. Cesium atoms are comparatively heavy and hard to get moving particularly fast, which means that the Doppler effects of lighter, faster particles are further eliminated. Their slower speeds also meant that the atoms would interact with the device longer, allowing a greater success rate at flipping the electrons to the proper state and siphoning out the ones that were in the incorrect state. This all meant a sharper "focus" for the device, which meant the frequency of the light that interacted most strongly with the cesium atoms was better nailed down.

In essence, the design by Ramsey started with a beam of cesium atoms, which were first passed through a magnetic filter to pick out only the atoms whose electrons were in the proper spin state. The successful atoms then went through a tunnel known as the Ramsey cavity, where they were hit with the wavelength of light that would cause an electron to flip its spin back over. The beam then passed through yet another magnetic filter that chose only the atoms whose electrons were in the right spin state. The more atoms that hit the detector, the better tuned the frequency emitted into the Ramsey cavity and the more certain scientists were of the frequencies of the quartz crystal oscillators they had embedded in the device. The wrong frequency would fail to force-flip the electrons, and few atoms would hit the detector. Underneath all the bizarre quantum physics and magnetic gatekeepers, it was nothing more than a positive-feedback loop, and the result was unprecedented timing accuracy.

This cesium clock was featured in a 1953 *Ripley's Believe It or Not* cartoon, with the caption, "It is even more accurate than the movement of the Earth!" Which isn't saying much, actually, because Earth's rotation rate can be affected measurably by large earthquakes. In 2011, a massive tsunami-spawning earthquake off the coast of Japan shortened Earth's rotation rate by 1.8 microseconds, not enough to affect your normal day-to-day activities, but enough to cause problems in guiding a spacecraft millions of kilometers to a precise landing on Mars. By 1967, cesium had become the basis for the standard definition of the second, which is now declared to be "9,192,631,770 periods of the radiation corresponding to the transition between the two hyperfine levels of the ground state of the cesium 133 atom."

The original atomic clocks were almost as cumbersome as the original computers were, but, unlike computers, which have been shrinking since their origin, atomic clocks grew in size for the first decade. The reason for this growth was that the longer the Ramsey cavity, the higher the accuracy. Their increasing lack of portability made it difficult to use them in any experiment requiring such precision timing, but to test Einstein's relativity meant that physicists would have to get these things moving somehow. In 1957, the United States felt the sting of Sputnik, but once the territory of space was made accessible, the idea of using it to test something as fundamental as Einstein's theory leaped almost immediately to the minds of physicists. In 1959, a *Popular Electronics* article featured Lyons and a trimmed-down atomic clock. The idea was that two clocks—one ground based and the other sent into orbit—would be synchronized on Earth. Then according to relativity, the time dilation from the satellite's speed would outweigh the shortening of time from the satellite's altitude. Once the satellite hit an altitude of 3,200 kilometers (2,000 miles), though, the gravitational effect would supersede the velocity effect. Lyons figured that the orbiting clock's information could be radioed to Earth and the times checked periodically. It seemed foolproof.

It never happened. Instead physicists decided to hitch a ride on something a bit more down-to-earth.

6   AROUND THE WORLD IN 80 HOURS (GIVE OR TAKE)

In 1971—16 years after Einstein's death—the definitive experiment was finally carried out. It required not a rocket launch but eight round-the-world plane tickets that cost the United States Naval Observatory, funded by taxpayers, a total of $7,600. The brainchild of Joseph Hafele (Washington University in St. Louis) and Richard Keating (United States Naval Observatory) were "Mr. Clocks," passengers on four round-the-world flights. (Since the Mr. Clocks were quite large, they were required to purchase two tickets per flight. The accompanying humans, however, took up only one seat each as they sat next to their attention-getting companions.)

The Mr. Clocks had all been synchronized with the atomic clock standards at the Naval Observatory before flight. They were, in effect, the "twins" (or quadruplets, in this case) from Einstein's famous twin paradox, wherein one twin leaves Earth and travels nearly at the speed of light. Upon returning home, the traveling twin finds that she is much younger than her earthbound counterpart. In fact, a twin traveling at 80 percent the speed of light on a round-trip journey to the Sun's nearest stellar neighbor, Proxima Centauri, would arrive home fully four years younger than her sister. Although it was impossible to make the Mr. Clocks travel at any decent percentage of the speed of light for such a long time, physicists could get them going at jet speeds—about 300 meters (0.2 mile) per second, or a millionth the speed of light—for a couple of days. In addition, they could get the Mr. Clocks out of Earth's gravitational pit by about ten kilometers (six miles) relative to sea level. And with the accuracy that the Mr. Clocks were known to be capable of, the time differences should be easy to measure.

This particular experiment had a gauntlet of computational complications, though, and Hafele attempted to account for all of them by imagining the experiment unfold from a great, stationary distance. For one, an altitude of ten kilometers is not much. Gravitationally speaking, Earth's surface is actually 6,371 kilometers (3,958 miles) from the center of mass, and different airports are at slightly different altitudes. Another ten kilometers turns out to make very little difference. On top of that, Earth is not simply a stationary body but is spinning on its axis (fortunately, Earth and its associated objects are all free-falling around the Sun, so its orbital motion becomes irrelevant in this experiment). This rotation forces everything on the planet to be in an accelerating frame of reference. Even the stationary ground-based clocks are not really stationary unless they're at the poles, so the clock used to synchronize the others experiences its own relativistic effects. Also, any airplane traveling eastward relative to the ground is actually traveling much faster absolutely than an airplane traveling westward at the same speed relative to the ground. Other problems arise from the fact that airplanes cannot simply fly completely around the world at a cruising altitude of ten kilometers. They have to land, refuel, and drop off and pick up other ticket-holding passengers. Quite often, in fact: on their trip around the world, they landed and took off over ten times each. These were, after all, ordinary commercial airliners, not dedicated physics labs. Even time spent at the gate at a latitude different from that of the Naval Observatory clock in Annapolis, Maryland, would make a difference. Hafele carefully computed the tiny effects from motion (special relativity) and altitude (general relativity), effects that added up and sometimes even canceled out.

The Mr. Clocks were up to the task of measuring predicted time differences in the tens of nanoseconds, though, even when other factors (temperature differences or external vibrations, for instance) were accounted for. And so off they went—two around the world to the east and two around the world to the west. The westward clocks were expected, after all was said and done, to gain 275 billionths of a second compared with the standard in Annapolis. Their special relativistic contribution from speed should have been about a third of this total, the greatest effect on the westward flights being the quicker aging due to altitude. The eastward ones, on the other hand, should have lost 40 nanoseconds, as their speed would have stretched time out more than their altitude would have compressed it.

That was the theoretical computation, anyway.

Each trip took about three days, during which the attending physicist would continually check on the Mr. Clocks' health (and consequently sleep very little). There was some uncertainty, as each clock suffered its own "drift" in timekeeping. The initial result showed that the westward clocks gained an average of 160 nanoseconds, while the eastward ones lost 50 nanoseconds. Certainly these numbers were in the correct sense, but the westward figure was distressingly off. Some computational gymnastics attempting to account for each clock's drift put the numbers at 273 nanoseconds and 59 nanoseconds, more in line with Einstein's prediction.

Again.

The general uncertainties in the famed Hafele-Keating experiment were troubling, though, so physicists devised more and more tests to verify relativity. Atomic clocks, which had experienced quite a growth spurt in their early years, ultimately became more and more portable. They also became more and more precise. These days, with state-of-the-art laser-driven aluminum ion clocks, even a 33-centimeter (one-foot) difference in altitude will produce a measurable difference in the rate of time. That's tabletop physics at its most bizarre, something Einstein would have been delighted to see. But he already knew that it would play out in this fashion. Once, when asked what he would have done if the Eddington expedition had failed to observe the bending of light, he stated, "Then I would have felt sorry for the dear Lord—the theory is correct." Adding layers of bizarre to the bending of light and the stretching of time is the current Gravity Probe B experiment, which shows that Earth's rotation drags space-time with it like a giant honey dipper twisting the thick, sticky contents of Winnie-the-Pooh's pot.

Other than stroking the egos of physicists, though, what can relativity do for anyone? It's one thing to joke that children simply age slower than grownups because they're closer to Earth, but another to make any practical use of the nanosecond consequences of our flexible space-time. In a single billionth of a second, though, light can go an entire foot. In the 275 billionths of a second accumulated by the three-day Hafele-Keating experiment, light could travel the length of a football field. But what about things moving much faster than jets? At much higher altitudes? Over time, their clocks would be less and less synchronized with clocks on Earth.

For the most part, this lack of agreement doesn't make a bit of difference to your average citizen, but it does if your average citizen is on the surface of Earth expecting a satellite more than 20,000 kilometers (12,400 miles) up and moving four kilometers (2.5 miles) per second to treat time the same way we ground-based creatures do. This is exactly why GPS satellites need to account for this difference. Initially conceived during the space race, our global positioning system would be pointless were it not for the fact that each space vehicle contains its own portable atomic clock. This time it's not to verify Einstein's relativity but to incorporate it. At the speeds and altitudes of these satellites, time would tick along nearly 39 millionths of a second faster each day. In a single microsecond, light can travel 300 meters (0.2 mile), so the 39-microsecond error from a mere day in orbit would mean confusing the position of an earthbound target by more than ten kilometers. To keep this from happening, the clocks on board the space vehicles are adjusted so that their frequencies are ever so slightly smaller (about one part in two billion) than their counterparts on Earth. Now every member of the small armada of GPS satellites sends out its signal 24/7 telling anyone who will listen what the exact time is. Your receiver picks up several of these signals, does a bit of calculating, and figures that if satellite A says it's this time, and satellite B says that it's this other time, and satellite C says it's some third time, this must mean that you are at the intersection of Gray Street and 17th Avenue.

If you have any sort of GPS system, you owe a debt of gratitude to Einstein and to the decades of physicists who worked, very often on the public dime, to create portable, highly precise clocks, whose oscillators were not springs or pendulums but atoms themselves. The century-old thought experiments of a curious character led to ever more rigorous tests of bizarre predictions, and those led to an industry that is expected to be worth over $26 billion by 2016. It has saved untold hours of time that would have been spent driving around aimlessly. It has led people out of danger, saved injured hikers, and helped motorists find the nearest gas station when they were running on fumes. Parents use it to track their children, receiving a text message if the baby sitter has taken the child beyond a certain distance. In 2013, German researchers showed that GPS could better map actual shifts in Earth's crust and provide life-saving minutes of warning for those in the path of a deadly tsunami.

It has become so commonplace, in fact, that the idea of getting a new cell phone without GPS capabilities is virtually unheard of. Why would you? It costs almost nothing these days, and it is practically guaranteed to be useful, if for no other reason than finding the nearest restaurant on your next road trip.

✸ ✸ ✸ Now imagine a world where GPS doesn't exist in everyday devices. A world in which you figured a couple of hours out of the house with your ailing child would not make a difference in his life except to cheer him—and you—up. And it does, until you return home to the devastating message that your child has been passed over for a heart transplant because nobody could find you in time.

PART II

✳✳✳ Identity Crisis

IMAGINE. In what must be a surreal nightmare, you find yourself standing before a judge as a jury foreman pronounces the word "Guilty." It echoes in your mind. Here you stand, an innocent person, and yet you've been found guilty of murder. Guilty of robbery. The victim was your next-door neighbor. The damning evidence? O-type blood—the most common blood type in the United States—at the scene of the crime and the word of your ex-girlfriend. Your fingerprints do not match those at the crime scene, but for some reason neither that fact nor your time cards from work are presented at the trial. Instead your life is quickly turned inside out, and you hear but cannot seem to comprehend the sentence. Life in prison. The years crawl by as you fight and appeal, hoping to break the stubborn walls of the criminal justice system. You learn of the fledgling science of DNA testing, but it remains too fringe, too untrustworthy to qualify as hard evidence. The irony does not escape you as you spend most of your adult life behind bars. Finally, after 17 years, the system has begun to trust the results of DNA testing, and the blood from the crime scene is tested against yours. The findings are conclusive: no match. You are free, as you should have been from the start. You've read and heard enough to know that you are free because forensic scientists were finally able to peer into the details of your genetic code, a code that, unlike your blood type, is uniquely yours. What you don't realize is that pulling out that information from a complex, microscopic molecule was made possible through a legacy of curiosity, some spare DNA in a microbiology lab, and an ecologist's failed attempt to grow a culture of the pink slime living in the hot springs of Yellowstone National Park.

7   ONE STRAND

You've got your mother's eyes. You're the spitting image of your great uncle. Curly red hair runs in the family.

You've probably heard these sorts of things at some point in your life. So have your mother, your father, your neighbor, and your great-great-great-great-great-great grandmother. The ability to inherit traits like eye color, hair texture, facial features, and height has been known since time immemorial. But, despite being perpetually immersed in countless examples of inheritance, mankind has spent most of history grappling with the underlying mechanism behind that inheritance. Simultaneously unimaginably complicated and deceptively simple, the code of life refused to be pinned down until relatively recently.

For centuries, natural philosophers struggled with the evidence, providing nothing more than qualitative—and sometimes amusing—guesses. When we say "in his blood" or "bloodline," we subconsciously cling to the mistaken notion that something in our blood is transferred from generation to generation. In fact, this is just one of many instances where the universe seems to have been playing a game of irony with us. We now know that red blood cells contain no genetic information.

The invention of the first crude microscope in the 1590s held some promise that we could get to the bottom of the roll of the dice that seems to be played every generation. Awkwardly, with the new instrument came the odd "observation" that little men inhabited sperm cells, leading some to conclude that the uterine environment did nothing more than allow these cells to develop into full-sized humans. How children could wind up sharing features with their mother was never fully addressed. An opposing camp with just as much evidence asserted instead that future offspring were housed in miniature within the egg cells of the female and that the father's job was merely to tell the egg that it was time to grow. But, if either of these cells had a tiny human inside, that would mean that the tiny human had its own sperm or egg cells with an even tinier human inside, on down the line.

As you can tell, the early microscope was not the appropriate tool for sorting out the intricacies of inheritance. Simple statistics and a controlled experiment with macroscopic objects, however, shone the first real light on the situation. After literally millennia of observing traits passed down through generations, and even artificially selecting for certain desired characteristics in crops and livestock, humankind finally had in the person of Gregor Mendel someone who approached the question rigorously and systematically. Even then, he was not taken seriously.

Born Johann Mendel in 1822 to a rural farming family in what was then the Austro-Hungarian Empire, he distinguished himself early on as a promising student of the natural sciences and mathematics. Further irking his father, who still held out hope that his academically inclined son would someday take over the family farm, he entered the St. Thomas monastery in Brno. Monastic life at this centuries-old complex was unorthodox at best. With a thriving culture of music and scientific pursuits, the monastery appealed to the 21-year-old Johann. Upon entering, he took the name Gregor, by which he is most commonly known, and immersed himself in the life of enlightened monks. Predictably, this university-like culture did not particularly appeal to the local bishop, who had a number of issues with the scholarly monks, not the least of which was the lack of order in this Order.

Mendel's first assignment as a parish priest didn't go particularly well. Dealing with sick and dying people unnerved him, and, anyway, he had entered the monastery largely to avoid having to worry about money for the rest of his life, not to be a compassionate minister. It wasn't that he didn't care. It's that the whole situation was extremely awkward for him. Upon realizing that Mendel's gift apparently lay elsewhere, the monastery farmed him out to the local village as a science teacher. It seemed a great fit for him. He was a popular and apparently talented teacher, but surprisingly he failed the certification exam. To better prepare him for the test, the monastery shipped Mendel off to the University of Vienna, where his talents continued to shine in the classroom but hid themselves at exam time. Arriving late one term, he was begrudgingly accepted as the thirteenth student in a 12-student group of teachers-in-training. Unfortunately—or fortunately, as it turns out—his original math and physics professor, Christian Doppler, for whom the Doppler effect is named, fell ill. Taking his place was Andreas von Ettingshausen, whose ideas about something called combination theory Mendel eagerly soaked up, along with emerging ideas about evolution and botany. Nourished by mathematical relationships and findings on plant hybridization, Mendel's own ideas would start to take root.

Sadly, his test anxiety never abated, and in 1856, he failed his certification exam for the last time. Mendel had to resign himself to being essentially a permanent substitute teacher. What seemed to be a humiliating setback left him plenty of time to explore the traits of pea plants.

Back at the monastery, Mendel began to experiment with mice in an effort to find some rhyme or reason behind the rules of heredity. The bishop strongly discouraged this pursuit because studying heredity in mice would necessarily mean that he would be monitoring their breeding activities. In a *monastery*. And anyway, they smelled awful, so the mice were expelled. Mendel turned to the less odiferous pea plants, amused that "the bishop did not understand that plants also have sex." He obtained 34 distinct varieties of pea plants that were "true" breeds. That is, every generation would possess a particular trait that he was looking for, like purple flowers or green seeds. He spent two full years methodically crossing them, an incredible feat considering that it takes just a breath of a breeze to send pollen where he didn't want it. From specially planned hybridizations, Mendel eventually wound up with nearly 10,000 subjects for study. Then he spent several more years interpreting his abundant notebooks of data. Finally, in 1865, Mendel presented his ground-breaking findings at the February and March meetings of the Brno Natural History Society.

The reception was not quite what he had hoped.

Rather than applaud him for his groundbreaking findings, the scientific community largely brushed off Mendel's work for being overly mathematical, something biologists of the era simply were not. Even 40 years later, a scientific paper by Dutch botanist Hugo De Vries—the scientist who coined the term "gene" and who helped rediscover Mendel's work—began not with an abstract but with an excerpt from a poem:

> Vom Vater hab ich die Statur,
> Des Lebens ernstes Führen,
> Vom Mütterchen die Frohnatur
> Und Lust zu fabulieren.

> (From the father I have the stature,
> Earnest conduct of life;
> From mother, good nature,
> And desire to imagine.)

> *J. W. Goethe, 1871*

De Vries' paper—"Fertilization and Hybridization"—has no equations, no tables, and very few numbers. In a field where qualitative and poetic discussions were at home in scientific conferences, Mendel's algebraic expressions and calculated ratios were not just out of place; they were unwelcome.

All of that mathematics revealed an amazing intuition on Mendel's part, however. For each of the seven characteristics he chose to study (e.g., wrinkled versus smooth peas, yellow versus green pea pods, tall versus short stems) he assigned a letter. He then created a sort of shorthand code for each type of plant. A specimen with wrinkled peas, yellow pods, and tall stems might be coded as aBC, while one with smooth peas, yellow pods and short stems would be ABc. Little did Mendel (or anyone else, for that matter) know, our genetic code, describing not only all of our physical characteristics but also some of our behavioral dispositions, would later be represented by sequences of letters. Billions of letters.

In the United States these days, a high school biology class would be incomplete without having students draw Punnett squares, which are efficient ways to visualize what Mendel reported in a far more cumbersome fashion in his 1865 presentation. It took 40 years, but clarity was finally achieved when Reginald Punnett first presented his genetic tic-tac-toe boards—now seen by ninth graders across the nation. Punnett himself was a great fan of Mendelism, even writing an entire book by the name. But he had struggled with rules of heredity, specifically why dominant traits didn't ultimately wipe recessive ones out of existence. Fortunately, he had a good friend and cricket teammate in the person of Godfrey Hardy, a mathematician. With his help, Punnett developed the visually simple idea still used over a century later.

Let's say you start with purebred plants with yellow peas and mix them with purebred plants with green peas. Following Goethe's poetic reasoning, every resulting plant should have a contribution from the yellow parent (Y) and a contribution from the green parent (G). As Mendel found out, despite having one parent that was a purebred green pea, all the offspring had yellow peas. Probably as important is the fact that none of the plants were any sort of yellow-green compromise. However, if you then mix the second generation pea plants—all with yellow peas, mind you—you find that green shows back up in a quarter of the resulting offspring. That is, assuming you perform enough crosses. Mendel's great strength lay in the sheer numbers of plants he crossed, allowing him to draw sound statistical conclusions. Punnett's strength lay in charting it out so well that the average 15-year-old can see what is happening. For the first cross, the parents are represented by the Ys along the top and the Gs along the side. The resulting offspring are represented by the YGs in the following Punnett square:

	Y	Y
G	YG	YG
G	YG	YG

All offspring could be represented as a combination of Y and G, instead of either completely Y or completely G. But since all the offspring appeared yellow (Y), Mendel claimed that there was something dominant about the yellow pea types that kept the green color from manifesting itself when the yellow component was present.

The second cross was where things became more interesting. If you start out with all YG plants, the new Punnett square looks like this:

	Y	G
Y	YY	YG
G	YG	GG

Thus a quarter of the offspring would be pure Y again (yellow-looking), while half would be the hybrid YG (also yellow-looking). What astonished Mendel was that a quarter of these third-generation plants would be pure G, showing green peas even though none of their parents had green peas. Mendel presented this result as a quadratic equation and charts of statistics, but the Punnett square makes it seem so obvious. What's more, Mendel suggested that some hereditary unit—possibly a particle—was passed from each parent to its offspring. Perhaps he failed to see his own conclusions clearly enough, or perhaps biology was not yet ready for him. After the cool reception, Mendel returned to monastic life and was soon elected abbot of St. Thomas Monastery, where he became more concerned with political and ecclesial matters and less concerned with scientific pursuits. Ironically, Mendel's findings were recessive traits in the scientific world, expressing themselves only after a generation of lying dormant.

8   TWO STRANDS

The latter half of the nineteen century was an incredibly important time, scientifically speaking, even if all the players were largely unaware of the significance of their parts. As Mendel's treatise on plant hybridization and all that it implied began gathering dust in academic libraries the world over, another apparently unrelated study into the chemistry of cells was revving up to be largely underappreciated. Born in 1844, Johann Friedrich Miescher was one of the pioneers in the budding field of physiological chemistry, or understanding the elemental makeup of living things. Against his father's wishes that he do something with immediate practical application, the Swiss-born Miescher found himself attracted to the bigger-picture questions. He was not so much interested in the practice of healing as in understanding the underlying mechanism of the process. He was intrigued by the obviously important role of lymphocytes, but, failing to extract significant amounts of those from blood samples, he turned to the generic category of white blood cells, or leukocytes. By the age of 24, he was working in the laboratory of renowned physiological chemist Felix Hoppe-Seyler at the University of Tübingen in Germany.

Hoppe-Seyler had done considerable research on the composition of hemoglobin, the molecule responsible for getting oxygen from your lungs to the rest of your body. Hemoglobin was easy to obtain in large quantities, as it constitutes a considerable percentage of the total makeup of blood. As Miescher soon discovered, finding a suitably large supply of white blood cells was a bit more challenging. White blood cells contribute only about 1 percent of the volume of blood, but conveniently they experience a surge in concentration around sites of infections. A testament to his determination and strong constitution, Miescher obtained his leukocyte samples from the local surgery clinic, which always had a stash of freshly discarded, pus-covered bandages that would instantly be declared biohazards these days. To separate the leukocytes from the cloth, he carefully washed the bandages in a solution of sodium sulfate, laboriously checking the microscopic behavior of the cells for each different concentration. A methodical chemist, he subjected the cells to various solutions to gauge their effect, sometimes adding a vigorous manual shake for good measure. The gel-like interior of the cells, or cytoplasm, was determined to be a combination of fats and proteins, something that had been studied in some detail by the physiological chemists of the era.

But it was the nucleus that most fascinated him. When the nuclei were subjected to a slightly acidic solution, an odd cloudy substance would sometimes rain to the bottom of the experimental vessel. It was found to be slightly acidic, and something that behaved like a protein, but not exactly. It was something new and captivating, puzzlingly rich in phosphorus and lacking in sulfur, but also consisting of the more familiar elements of life: hydrogen, carbon, oxygen, and nitrogen. Since it came from the nucleus itself, he dubbed it "nuclein."

Further exploration into the nature of this substance required a large dose of creativity on Miescher's part. Wanting to ensure that stray proteins didn't interfere with his analysis of the chemical composition of this strange nuclein, he employed a method that has been used by Mother Nature to get rid of them: he digested them. Not personally, of course, but he did put the pus washed from the bandages into a solution derived from the stomach of a pig. Pepsin, he hoped, would rid the cell completely of stray protein impurities by breaking them down, leaving only the pure nucleus of each cell for analysis. His experimental setup in Tübingen Castle in Germany looked for all the world like a still. However, the intoxicant he sought was not ethanol but insight into the chemical process of life. Since nuclei were present in so many types of cells, he decided to hunt down nuclein in places other than discarded surgical bandages. And he found it all over the natural world. For reasons that he could not support experimentally or even articulate, Miescher knew that in this white fluffy precipitate lay something monumentally important. "But why discuss possibilities?" he wrote in 1871. "The analysis of cells at different stages of development will surely give good clues."

He would wrestle with the purpose and makeup of nuclein, meanwhile dancing unbelievably close to the correct explanations for heredity, for the next 20 years, until finally the incessant hours in the lab took their toll. He contracted tuberculosis and retired to the Swiss Alps, where he would live out the last months of his relatively short 51-year life.

Other scientists set out on the same path, even reproducing his results. Nuclein came to be known, much to Miescher's dismay, as "nucleic acid," the N and the A of what we now know as DNA. Albrecht Kossel, another who cut his research teeth in Hoppe-Seyler's lab, ambitiously took Miescher's baton, discovering that nucleic acid consisted of four nitrogenous bases and sugar. His Nobel Prize–winning work describing the four bases provided the scientific community with four new letters: A, C, T, and G. These stood for adenine, cytosine, thymine, and guanine, the last one so named because it was also found in guano—bird (or bat) poop.

Ironically, because only four main bases made up this nucleic substance, it was generally felt that they couldn't possibly be the mechanism behind heredity. There was simply too much variation in the life on Earth to be accounted for by only four building blocks, and so the detective work continued.

9   THE FIRST RUNGS OF
    THE LADDER

Mendel's work remained not merely unheralded but literally unopened in many libraries around the world. Forty years later, copies of the conference proceedings still had uncut bindings. As for Miescher's achievements . . . well, just ask yourself if you have ever heard of him. And yet the contributions of these two perfectionists stand as the rails of the ladder of our understanding of heredity. One provided the theoretical background; the other found strong clues pointing to the physical mechanism. But the rungs between these two rails had not yet been built, leaving both unstable. Connecting them would require open minds, better microscopes, and dead mice. Lots of dead mice.

Vastly improved over the original, almost accidental, crude models that afforded perhaps a magnification of 50 times, the nineteenth-century microscopes were designed not by trial and error but around equations and theories about the limitations of optical systems. German lensmaker Carl Zeiss began creating exquisite microscopes in the mid-1800s, and the company that bears his name is currently one of the largest and most well-respected manufacturers of high-end camera lenses and microscopic equipment.

These new, high-powered, clear views of the microscopic world greatly aided our understanding of the mechanism of life. Once discernible only as a small unit of a much larger organism, an individual cell was now seen to be a thriving metropolis, and the nucleus was clearly an important hub of activity. The cellular building blocks of life could even be witnessed dividing, not on timescales unfathomable to humans but over the course of minutes or hours. The patient observer could watch as parts of the cells organized themselves, copied themselves, and finally pushed the cell apart into two separate but apparently identical "daughter" cells.

The discovery of the intermingling of actual particles from each parent during reproduction was a surprising spectacle that the new and improved microscope afforded scientists. Even into the mid-1800s, the debate had raged regarding the actual purpose of these cells, and the idea of a tiny person (or at least the essential parts for one) inside the egg or sperm cell endured. The fact that the sperm cell could be personally witnessed entering the egg was heralded as a victory for both sides of the preformation debate. Clearly, the spermists agreed, all the parts required to make a person are using the egg merely to set up shop. Clearly, the ovists agreed, all the parts required to make a person are already inside the egg and the sperm are initiating the assembly and growth of these parts. Even when the microscope provided visual evidence contrary to these ideas, the reaction was often the scholarly equivalent of a young child plugging his ears and screaming, "I can't hear you!"

But in 1876, a decade after Mendel's publication on the hereditary patterns of pea plants and seven years after Miescher's discovery of nuclein, Oskar Hertwig reported unequivocally that the nucleus of each type of reproductive cell (at least in sea urchins) actually fused into a single nucleus during fertilization. It was not one or the other. Both were necessary and complementary.

Another of the more stunning revelations to come from microscopic studies was the intricate choreography of strange filaments within the nucleus. At a time that the cell appeared simply to "know" to be right, the boundary of the nucleus would dissolve, and the once-chaotic tangle of spaghetti-like strands would gracefully morph into a number of shadowy forms that often looked like abstract butterflies, sometimes with clear variety in size and each one with two seemingly identical wings. Then, continuing the dance, all the tiny butterflies would line up parallel to each other along an invisible line, where the choreography took a shocking turn. The cell would appear to play tug of war with each butterfly, pulling its wings toward opposite poles. All the left wings would fold up and gather to the left to form an odd mop-head shape; all the right wings would gather to the right in another mop-head that mirrored the leftward one. Each set of wings would then be enveloped by its own nucleus, and the cell itself would be pinched in the middle by an invisible hand. Two cells where only one had been. The spaghetti-like strands would revert to a chaotic mess within the nucleus, and at the appropriate time the process would begin anew.

"Omnis cellula e cellula" ("All cells come from cells") declared German polymath Rudolph Virchow in 1858 after first witnessing this division. Not only that, but everything that the nucleus once contained was now contained in duplicate, leading Walther Flemming to provide the 1880 addendum: "Omnis nucleus e nucleo" ("All nuclei come from nuclei").

The "butterflies" and other shadowy symmetric shapes were dubbed chromosomes, derived from a Latin term meaning "colored body," so named because, luckily for early microscopists, these molecules readily absorbed dyes and were easy to spot in cells. The microscope revealed other telling processes, though. Not all cells had the same complement of chromosomes. In 1901, it was found that the reproductive cells had only half as many as other cells in the body. Only upon the union of the egg and sperm would a full complement that owed half its identity to the mother and half to the father arise.

It was the American Walter Sutton who, at Columbia University, first linked the behavior of the very small chromosomes to the visible outward signs of heredity. In his 1902 paper discussing grasshopper reproduction, he asserts, "I may finally call attention to the probability that the association of paternal and maternal chromosomes in pairs and their subsequent separation during the reducing division . . . may constitute the physical basis of the Mendelian law of heredity." Hot on the heels of this paper, Sutton published "The Chromosomes in Heredity," where he revived Mendel's lettering schemes for traits. He drew a stunning conclusion. If an organism has 16 chromosomes, or eight pairs of chromosomes, then there are 256 (or two to the eighth power, eight being the number of chromosome pairs) ways of combining those into a single reproductive cell, or gamete. And if that gamete meets the complementary reproductive cell, also with 256 possible combinations for its chromosomes, then the resulting offspring can have as many as 65,536 chromosomal combinations. Perhaps, for Mendel's pea plants, it could be a tall plant with smooth, yellow peas or a short plant with smooth, yellow peas, or . . . the list of possible trait combinations was extensive, which is why only an experiment like Mendel's, involving thousands of different crosses, could ever have made any sense of them.

Sutton was beside himself. "It is this possibility of so great a number of combinations of maternal and paternal chromosomes in the gametes which serves to bring the chromosome-theory into final relation with the known facts of heredity."

The puzzle was slowly coming together.

10   INTERCHANGEABLE PIECES

Those chromosomal combinations seemed to imply that there were pieces to the genetic puzzle that could be swapped out and replaced with others. Pull out the piece for "yellow" and insert "green," and you have a new pea plant. Life seemed to choose from a menu of interchangeable options like a child decides which Lego blocks to build her dinosaurs with. Black then blue then yellow? Or black then red then white? Although microscopes in the early decades of the twentieth century could not detect activity like this on the molecular level, such swapping could be inferred if only we could pick an "option" and allow nature to insert it.

Tragically for the mice involved, the option their human observers chose was death. It was 1928, and Fred Griffith reported on his exhaustive six-year-long study of the effects of different strains of the *Streptococcus pneumoniae* bacterium on mice. Griffith was a bacteriologist by training, meticulous in the spirit of Mendel and Miescher, and researching an illness that took the lives of many: pneumonia. He was working in the Pathological Laboratory of Great Britain's Ministry of Health, a government worker with the honorable practical goal of understanding the mechanism by which pneumonia killed. In his experiments, he sought answers to the bacteria's virulence, specifically why it killed some people but not others.

The particular type of bacteria that he studied comes in two main varieties: a type that built a smooth-looking colony and one that assembled into a colony with a rough appearance. The smooth strain appeared as such because of a protective polysaccharide coating that effectively kept the warriors of an attacking immune system at bay.

Once ensconced, these bacteria could multiply, ultimately killing the unfortunate host. Such had been the case for the involuntary subject who donated the lung fluid for Griffith's cultures after she had succumbed to pneumonia. The rough strain, in contrast, had no such coating, its colonies appearing ragged and uneven. This particular strain was not an automatic death sentence.

From 1922 to 1927, Griffith cultured the different strains and tested their effects on hundreds of luckless mice. One particular smooth strain killed virtually all the mice, sometimes within hours. The rough strain in mice, just as in people, was not often fatal. The seemingly blessed mouse receiving an injection of the rough bacteria would live, only to meet its untimely end at the hands of its human experimenter, who would watch the mouse for upward of three weeks before killing it (most likely by decapitation or by swiftly breaking the mouse's neck, or so modern-day biologists tell me) and analyzing its blood.

Heating the fatal strain, however, appeared to render the bacterium harmless. Griffith heated his cultures to varying temperatures for varying times to find the breaking point of the smooth strain. Once killed, the bacteria injected into the mice had no effect, and those mice would instead later die from unnatural causes, not pneumonia. This result made complete sense. After all, why should a dead bacterium cause any harm?

What did raise Griffith's eyebrows was the fact that simultaneously injecting both a live version of the rough strain (harmless by itself) and a dead version of the smooth strain (also harmless by itself) would result in a dead mouse. Moreover, the culture derived from the dead mouse's blood would show the smooth colonies, although all the smooth-colony bacteria had been killed.

His first thought was the most natural one: something had contaminated the sample and some smooth-strain bacteria had managed to get to the mice. But, once he carefully ensured that this was not the case, he had to reach a strange conclusion. The message of "smoothness" was somehow infiltrating the rough-colony bacteria's very identity. It was, according to Griffith, a transforming factor, something that was not destroyed by the same temperatures that killed the bacteria itself. A molecular "Lego block" that coded for smoothness was being picked up from the remains of the dead bacteria and inserted into the instructions to build the rough strain.

Again, the reception to this game-changing discovery was cool at best. Critics initially accused Griffith of running a sloppy experiment, but they were easily able to replicate his experiments and, more important, his conclusions. For his own part, he never really followed up on the nature of his "transforming factor." Its significance, like that of so many discoveries, would not be fully known for yet another generation. Instead, he continued working in the Ministry of Health, publishing little and yet steadily contributing to our understanding of various deadly diseases. He died on April 17, 1941, a night when the German forces bombed London and, tragically, the government laboratory where he and a colleague were working late. A shy, unassuming gentleman, he never saw his 1928 finding for what it was: "a delayed-action fuse which, 25 years after its publication, triggered off an explosion of biological knowledge, comparable only to that ignited . . . by the work of Mendel."

11   IGNITING THE FUSE

The story of genetics has taken us through a Czech monastery, a German castle, a university in New York City, and a British government laboratory. It has looked at outward signs of heredity and the inner workings of cells, tidy statistics and bucketloads of dead mice, the results communicated across generations, languages, and political borders through the standard of scientific communication: journal articles. For decades, progress was made erratically, sputtering along, then surging forward, fueled by findings that lay dormant until the scientific environment was receptive. In short, it was looking like an abbreviated version of evolution itself.

Despite the apparent lack of appreciation for Miescher's work, scientists began to study nucleic acids in great detail, and by 1930, two types had been identified. The first came from yeasts and contained the typical nitrogenous bases of adenine, guanine, and cytosine. However, instead of thymine, this type of nucleic acid contained uracil. In addition to the bases, there was a phosphoric acid component along with a sugar component: ribose, indicated chemically as $C_5H_{10}O_5$, meaning that it was a structure consisting of five carbon atoms, ten hydrogen atoms, and five oxygen atoms. Another type of nucleic acid possessed four bases that included thymine, along with a phosphoric acid component and a sugar dubbed deoxyribose. As its name suggests, deoxyribose has one fewer oxygen atom per molecule than ribose, and it forms the spine of the nucleic acid that, as far as anyone could tell, came from animal tissues. Scientists toyed with the name zoonucleic acid to indicate as much, but, since there seemed significant overlap in the types of nucleic acids found in plants, animals, and fungi, they went with the more generic descriptor: deoxyribonucleic acid (often called desoxyribonucleic acid in the early papers). This moniker was essentially cemented in 1931 when Phoebus Aaron Levene and Lawrence Wade Bass published their definitive tome, "Nucleic Acids," including within it the familiar letter triplets DNA and RNA (ribonucleic acid).

The question on the minds of many scientists was where the transforming factor that Griffith had stumbled onto was stored. Once that was figured out, biologists would then know where the key to heredity lay. Griffith had died with the common assumption that hereditary factors were managed by the proteins, but by the 1940s, it came down to a simple, logical process of elimination. It could be something in the DNA itself, the RNA, or the proteins. All it would take is carefully destroying those three potential hiding places and checking to see if transformation from rough *Pneumococcus* bacteria to smooth ones still took place.

The decisive work was carried out by a trio of scientists at the Rockefeller Institute of Medical Research, founded in 1901 by John D. Rockefeller Sr. Just months before establishing this institute, the eldest Rockefeller had witnessed his three-year-old grandson succumb to scarlet fever, a common *Streptococcus* infection that is effectively beaten by readily available antibiotics today. Hoping to help others avoid such heartbreak, he donated $200,000—equivalent to over $5 million today—to provide grants to medical scientists so that they could explore causes and cures for devastating diseases like tuberculosis, diphtheria, and typhoid fever. It was the first grant of its kind in the United States.

One of the shining stars of the Rockefeller Institute was Oswalt Avery, who joined in 1913 and was a nearly 30-year veteran there when he teamed with Colin MacLeod and Maclyn McCarty to get to the bottom of the transforming factor mystery. With decades of exploring DNA, RNA, and proteins, biochemists had been able to find out not only their basic composition but also their undoing: the "-ases." While some "-ase" molecules help build things up, other "-ase" molecules tear them down. Protease, for instance, is a class of enzymes—an "-ase"—that signals the downfall of proteins. Your body is swimming with various proteases, from the pepsin in your stomach that disassembles proteins in your food to the plasmin in your blood that helps regulate clots. There are even proteases in laundry detergents to help eradicate tough protein-based stains. Ribonuclease (RNase) and deoxyribonuclease (DNase) are, following the naming conventions, enzymes that break down RNA and DNA, respectively.

All that Avery, MacLeod, and McCarty needed to do was to perform Griffith's experiment but add a twist by systematically destroying one of the three suspects: the proteins, the RNA, or the DNA. If the proteins were destroyed and the smooth colonies arose from the dead, then proteins could not be the seat of heredity. Likewise, if the RNA was destroyed and smooth colonies of bacteria were found, then RNA was not the answer, either. What they found was that only DNase prevented the transforming factor from crossing into the rough colonies. Their 1944 conclusion was stated simply: "The evidence presented supports the belief that a nucleic acid of the desoxyribose type is the fundamental unit of the transforming principle of *Pneumococcus* Type III."

Like it or not, DNA was apparently the source of heredity.

In 1944, though, this was a rather unpopular finding, as most in the field had placed their bets on the proteins. There were, after all, many more amino acids—20 of them, to be exact—in biological proteins than there were nitrogenous bases in DNA, which sported only four basic building blocks. It was inconceivable that the diverse books of life could be written with only four letters. Logically speaking, the denial made no sense. It was almost like appearing on a game show where there are three doors and you are given three chances to find a prize. You open all the doors, seeing nothing behind the first two, but you still close your eyes to the prize behind door number three. Perhaps, they thought, if we just keep opening door number one in different ways, the prize will appear. Perhaps the prize behind the first door was momentarily disrupted by opening the third door. It was a deeply held conviction that took many different experiments of greater complexity to unseat. But that door was shut permanently in 1952, when Alfred Hershey, Martha Chase, and a kitchen blender definitively showed that proteins were not the answer. (I leave it as an exercise for the reader to find out how two scientists and an Osterizer proved once and for all that DNA was the seat of heredity.)

Fortunately, the structure of DNA, whether or not it was proven to be the key to inheritance, had been rigorously explored in the meantime. The molecule itself was a captivating puzzle. By the late 1930s, scientists had found that it had a shockingly high molecular weight. Furthermore, DNA was extremely long compared with its width. At the University of Leeds in England, William Astbury had attempted studying the structure of the molecule by shining x-rays on it, a trick that had been useful since 1914, when the precise layout and relative spacing of the atoms in table salt was divined. Because of the way the tiny wavelengths of x-ray light interact with the individual atoms in a molecule, shining x-rays on a substance with a regular structure, such as a crystal, produces a distinct pattern that provides valuable clues about how the atoms are arranged. Similarly, when x-rays are shone on a molecule of DNA, hints about its structure are revealed, although the complexity of this molecule makes a detailed analysis impossible. When Astbury did just this, he found that, at least viewed from the side, DNA has portions that are stacked one on the other "like a pile of pennies."

It's fascinating to wonder how scientists would have treated Astbury's "photograph" had they fully understood that DNA was more than just, perhaps, a scaffolding molecule. As it was, DNA was little more than a curiosity in the 1930s. Only after Avery, MacLeod, and McCarty had implicated it as the molecule of heredity did its structure really grab the attention of scientists around the world.

In 1952, with the full understanding that DNA held the secrets to genetics, the molecule was once again explored using x-rays, this time with greater detail in the resulting image. The photographer was Raymond Gosling, a graduate student of physical chemist Rosalind Franklin. It's likely that neither of these names rings a bell with you, but it's somewhat more likely that the names Watson and Crick do.

Ironically, neither name associated with the discovery of DNA's structure originally set out to find it. The Englishman Francis Crick was working on his dissertation studying not DNA but hemoglobin, the molecule in red blood cells responsible for transporting oxygen to every corner of the body. His future research partner, James D. Watson, was an American zoologist and geneticist whose interest in DNA came only after hearing physicist Maurice Wilkins give a presentation on the use of x-rays to probe the mysterious molecule. Watson left Chicago and eagerly joined the x-ray crystallography team across the ocean at Cambridge's Cavendish Laboratory. In England, he convinced Crick to abandon his dissertation on hemoglobin—a remarkable feat, considering how attached many graduate students become to their dissertation topics—and join him and Wilkins in the pursuit of DNA's hidden secrets.

For her own part, Rosalind Franklin had been purposefully working with Wilkins to unmask the secrets of the DNA molecule using x-rays, a technique for which she had become a respected authority. But theirs was a tense and often contentious professional relationship. Her student, Gosling, would later recall that "they just did not mix from the word go . . . they never sat down together long enough to exchange ideas." In fact, the professional climate at King's College where she and Wilkins worked practically guaranteed that she would never casually bounce ideas off him or any of her other colleagues. At that time, women were prohibited from using the faculty dining hall, the de facto venue for intellectual free association and inspiration.

Fed up with conditions that would be grounds for a modern-day lawsuit, she left, but not before Gosling had created what would become famously known as "Photo 51." This x-ray image revealed a pattern of small, shadowy rectangles creating a larger X shape. The rectangles became noticeably fainter and more diffuse the farther from the center they were. It was a strange, abstract image encoding an incredible wealth of information for those who could interpret it correctly. It is unclear whether Franklin would have drawn the appropriate conclusions, but what is clear is that Photo 51 made its way from her lab, which she had entrusted to Wilkins in her absence, into the hands of Watson and Crick. Having struggled with a physical model of the DNA molecule, particularly how it could make copies of itself, they saw everything they needed to see in that shadowy photo. DNA was a twisted ladder—a double helix.

But the battle was not over. How did the rungs of the ladder line up so perfectly without creating havoc in the orderly helical structure? It was not a stunning leap of physical insight that helped Watson and Crick here but instead a technique worthy of a fifth-grade science project: they used cardboard cutouts to model it. Once they aligned the molecules making up the rungs in their proper configurations, they saw that the puzzle fit together neatly. The two main backbone strands wrapped around a common axis, connected by specific molecular pairs. The size and bonding properties of these pairs clicked neatly into place like well-designed Tinkertoys.

This told them something even more profound. At the end of their 1953 paper in *Nature* outlining their results, they state, "It has not escaped our notice that the specific pairing we have postulated immediately suggests a possible copying mechanism for the genetic material."

For ferreting out the underlying architecture of the molecule of heredity, Watson and Crick, along with Wilkins, were awarded the 1962 Nobel Prize for Physiology or Medicine. Tragically, Franklin died in 1958, making her ineligible for the award, but even if she had lived, it's unlikely that she would have received full credit for her contributions. Watson later included in his autobiography, *The Double Helix: A Personal Account of the Discovery of the Structure of DNA*, a scathing portrayal of Franklin, an opinion that softened as the years wore on. For her part, Franklin would be honored not with a Nobel Prize but with Google doodle, a whimsical tribute that often replaces the usual Google logo on that search engine's main site, on her birthday.

The collective efforts to understand DNA, though, have helped scientists unravel quite a bit about this amazing molecule. Buried deep within nearly every one of your cells is the instruction manual for you. If stretched out, a single DNA molecule would be less than a thousandth the width of a human hair and about as tall as you are. Yet, shockingly, this enormous molecule can coil up so tightly that it becomes vanishingly small, unobtrusively residing in the nucleus of a cell with room to spare, a property that the twisted structure and special folding characteristics allow.

Conceptually DNA is relatively simple. Too simple, if you recall, for the generation of scientists who refused to accept that something so uncomplicated could possibly encode the instruction manual for all living things. In its tightly wound, twisted ladder, DNA has rails of alternating phosphate molecules and sugar molecules. Those rails hold rungs consisting of pairs of the four bases whose identities were revealed in the 1920s. It turns out that each base attaches with only one other. Adenine, whose chemical formula is $C_5H_5N_5$, can link only to thymine ($C_5H_6N_2O_2$). Cytosine ($C_4H_5N_3O$) can pair only with guanine ($C_5H_5N_5O$). Thus the individual rungs will contain either an A-T (or T-A) pair or a C-G (or G-C) pair but no other combinations. What makes it possible for these two simple pairings to encode everything that makes up the physical "you," from your eye color to your tendency to collect abdominal fat to your predisposition to aggression to a higher probability of developing breast cancer, is the sobering fact that this ladder is more than three billion rungs tall. Rearrange the rungs and you create a unique being with different characteristics.

It is perhaps surprising that 99.9 percent of your genetic material is identical to that of every other human on the planet (and likely surprising that 70 percent of your DNA is identical to that of a sea sponge). It's the minuscule variations in the other 0.1 percent that create your unique identity, and even identical twins have been found not to be 100 percent identical, genetically speaking. Moreover, virtually every cell in your body has your unique DNA inside it. Leave so much as a stray hair on your desk, and you've left your genetic diary open for someone with sufficiently powerful technology to peruse.

First, though, somebody had to create that technology.

12   BREAKING DOWN, BUILDING UP

One of the catalysts of that technology came from the unlikely figure of Arthur Kornberg. The son of hard-working Jewish immigrants, he showed no particular early interest in science, although he did manage to skip three grades in the Brooklyn public school system. At an age when other budding scientists were already building homemade chemistry labs or pondering the speed of light, Kornberg admitted to collecting matchbook covers. Graduating in 1937 with a bachelor's degree at the early age of 19, Kornberg decided to go to medical school simply to maintain the psychological security of being in school. The biochemistry units were tedious; the pathology units convinced him he had a dozen fatal diseases. Only the anatomy sections held his attention as he marveled at the intricacies in constructing life. By all accounts he shone in medical school, but a stubborn culture of anti-Semitism meant that he was passed over for fellowships. So, off to the Navy he went. After a brief stint as a ship's medical officer, he found himself at the National Institutes of Health, where he became deeply involved in determining the nutritional requirements of rats.

It was a far cry from anything to do with DNA.

Except that it wasn't exactly. Rat nutrition depended upon a number of things, not the least of which was the mechanism by which nutrients were metabolized by the cells. It was fundamentally biochemistry, even if he had explored it at only the most superficial level. By 1945, Kornberg dug deeper into the world of metabolic enzymes.

The word enzyme itself means "in yeast," as the nineteenth-century insights into the importance of these molecules were often made while observing the process of fermentation—brewing and bread making, to be more specific. Enzymes are chemical machines, each with a specific job of building up or tearing down, and Kornberg's love for them is obvious in this tribute;

> What chemical feature most clearly enables the living cell and organism to function, grow, and reproduce? Not the carbohydrate stored as starch in plants or glycogen in animals, nor the depots of fat. It is not the structural proteins that form muscle, elastic tissue, and the skeletal fabric. Nor is it DNA, the genetic material. Despite its glamour, DNA is simply the construction manual that directs the assembly of the cell's proteins. The DNA itself is lifeless, its language cold and austere. What gives the cell its life and personality are enzymes. They govern all body processes; malfunction of even one enzyme can be fatal. Nothing in nature is so tangible and vital to our lives as enzymes, and yet so poorly understood and appreciated by all but a few scientists.

Their chemical structure turns enzymes into a molecular extra pair of hands. Just as you sometimes need one more hand to hold a piece of wrapping paper down while you put the tape on, molecules sometimes need a bit of extra help to do their assigned job. For instance, water molecules have the job of splitting the sugar sucrose into fructose and glucose. But they can't do that on their own. The enzyme sucrase is a pair of hands that hold the sucrose in such a way that the water scissors can cut it in two. Without the extra hands, the sucrose is too strong, too unwieldy for the scissors alone, but, once the help of the enzyme sucrase is employed, the cut can be made. Two new molecules—glucose and fructose—are sent on their way (the scissors in this case are split and absorbed into the new product molecules). The enzyme is unchanged by the encounter and moves on to help another water molecule cut sucrose apart.

For every vital reaction, there is a specific enzyme to help. Sometimes these enzymes subcontract to other molecules. Kornberg found out that some nutrients—thiamine, for example—are indispensable coworkers of enzymes. "I vividly recall my excitement in reading for the first time about these enzymes, coenzymes . . . I also recall my astonishment on hearing a seminar at NIH . . . demonstrating that one gene is the blueprint for one enzyme. I knew even less about genetics than about biochemistry," Kornberg related in his autobiography.

It's no wonder, then, that, when he first attempted to report that he had discovered the molecule responsible for assembling DNA, the reception was anything but welcoming. He had invested a mere decade in molecular construction and deconstruction, and DNA wasn't even his area of expertise. The original reviewer's comments to his 1957 submission to the *Journal of Biological Chemistry* cut to the chase, stating, "It is very doubtful that the authors are entitled to speak of the enzymatic synthesis of DNA."

Watson and Crick had assumed that DNA simply unwound itself naturally, allowing for clean, quick replication. Kornberg knew after a decade studying enzymes that this was highly improbable. If it takes an enzyme to help break a comparatively simple sucrose molecule, how likely would it be that the multibillion-atom complex that encodes all of life could take care of its own deconstruction and construction all by itself?

Kornberg admits that DNA was never his true focus, that he was not just jumping on the Watson-Crick bandwagon in the 1950s. He was simply studying his favorite molecules, enzymes, and going where they led him. But, when he found that there was an enzyme that cut a molecule into two nucleotides (a nucleotide is a base with its phosphate and sugar molecules, or half a DNA ladder rung along with its part of the ladder rail), his interest was piqued. Perhaps there was something that could join nucleotides, not separate them, something that could create the ladder from its component rungs and rails. He began work on hunting down enzymes that could not simply break down but bind up.

In 1953, Kornberg's success came as almost a byproduct. He had isolated considerable amounts of DNA for use in a microbiology class, and, since he had so much of it, he decided to go ahead and try to experiment on it. He capitalized on the discovery by Miescher that DNA rains to the bottom of an acidic solution and yet individual nucleotides float around within that same solution. He "tagged" those free nucleotides with a radioactive tracer. If he put those tagged nucleotides in a solution and they found their way into a DNA molecule, Kornberg would see radioactive DNA rain to the bottom when the solution was acidified. If they wound up inside the DNA, then something had to have put them there, and that something was the machinery that assembles DNA.

It turned out that Kornberg wasn't the only person exploring the role of enzymes in building up the molecules of life. At New York University, Severo Ochoa was succeeding in uncovering the enzyme responsible for building up and tearing down RNA: polynucleotide phosphorylase. It's quite a mouthful, but it was a shocking piece of news to Kornberg, who redoubled his efforts to synthesize DNA in 1955. Again he tagged nucleotides with radioactive tracers and introduced them into a solution with the presumed DNA builders. Acidifying the solution made the DNA rain down, and again he found radioactive tracers incorporated in the DNA. Using the knowledge gleaned by Avery, McCarty, and MacLeod a decade earlier, he used DNase to dismantle the newly synthesized DNA. The radioactivity was no longer found in the precipitate, meaning it really had been built into the DNA molecules themselves. Once the basic ingredients had been assembled and the steps determined, it was amazingly straightforward to isolate DNA polymerase.

For this discovery, Kornberg would first experience the wrath of journal editors who scoffed at his choice of name for the molecule and then his credentials. To even dare making such an announcement, he should have been "somebody." Just two years later, he was awarded the 1959 Nobel Prize in Physiology or Medicine, an honor shared with Severo Ochoa, who isolated the enzyme responsible for the creation of RNA, without which DNA would be like a script for a play that was never performed. But the stage was set, the actors awaiting their big break.

13 A CURIOUS BOY

While all of this drama was playing out in labs around the globe, there was a curious Ohio boy oblivious to most of these breakthroughs in genetics. Thomas Brock was not yet two when Griffith reported on the transforming factor, and, while Avery, McCarty, and MacLeod were starting to wrestle with the exact location of that transforming factor, Brock was blowing things up in his makeshift chemistry lab in the barn behind his house. By the time DNA was implicated as the molecule of heredity, Brock was fresh out of the Navy and on his way to becoming a writer. While an undergraduate at college, he found himself drifting back to the sciences. In 1952, as Watson and Crick were putting the final touches on their Nobel-worthy announcement about the structure of DNA, Brock was finishing up his PhD studying mushrooms and yeast, a far cry from anything to do with your genetic code, and looking for a job. With his newly minted doctorate in soil fungi, he found himself with limited academic job prospects, so soil fungi would have to wait, just as so many of his other pursuits had. However, there *was* a loosely related private industry that was hiring scientists who understood life on small scales: pharmacopoeia. That is, drug manufacturing.

By the time Brock was job hunting, the science of drugs had gone well beyond guesses, elixirs, and snake oils of the nineteenth century. Instead, treatments were established based on scientific investigation and biochemistry. The Upjohn Company of Kalamazoo, Michigan, had been a visible presence since 1886, when the biggest challenge facing drug makers was creating a pill that didn't simply pass through the body unabsorbed. Then, as now, Upjohn was unafraid to market itself, slowly gaining fame for its easily dissolvable pills and candy-based laxatives. Despite the gimmicks (Upjohn handed out bottles with candy pills at the 1893 Chicago World's Fair), the company strove to remain grounded in science.

Stimulating the entire pharmaceutical industry was Alexander Fleming's fortuitous 1928 discovery that *Staphylococcus* bacteria could not grow on bread mold. There was a rush to find out why, and, more important, if there was a way to isolate the bacteria-killing substance in medicinal form. Ultimately a moldy cantaloupe from a Peoria farmer's market held the key to the development of penicillin, a drug that was seen as a magic bullet aimed at Death himself. During World War II, countless wounded soldiers faced what would have been deadly infections were it not for the miracle of penicillin. Upjohn was selected by the U.S. armed forces to supply penicillin and other drugs to the troops, catapulting the already successful company to the forefront of the industry. By the time Brock was searching for a job, one of the more vibrant divisions of the still-growing company involved research into antibiotics.

Although not a bacteriologist, Brock was a quick study, and his knowledge broadened to include microbiology and, of all things, the German language, which he picked up in his spare time, because his curiosity was somewhat stifled in the corporate environment. Five years later, in 1957, he transitioned back into the world of academia, where unbridled curiosity was part of the job description. As he taught his classes, he taught himself. An expert in yeasts, he received a federal grant from the National Institutes of Health to study them. An expert in antibiotics after his stint at Upjohn, he received a federal grant from a nascent National Science Foundation to study them. Ever inquiring, he played with radioactive substances, training himself how to use them as biological tracers. He even translated his own book on microbiology, a subject he had taught himself, into German, a language he had taught himself. His love for learning never took a vacation, even when he did. He often joined his wife's family for camping and lake kayaking, activities that kindled in him an interest in aquatic life. These days, he might have been diagnosed with attention deficit disorder, but in the 1950s, such far-reaching curiosity was rewarded.

Part of the reason behind this welcoming climate was the United States' postwar economy. Manufacturing had taken off, and the suburbs sprang up with an explosion in population. Few people were concerned that their tax dollars were providing funding for a scientist to toy with radioactive tracers in yeasts or that he had abandoned what was clearly a more practical use for his intellect. In fact, pure research endeavors such as Brock's had been given the presidential blessing in 1950, when Harry S Truman signed the National Science Foundation Act. "Throughout our history, scientists and scientific knowledge have contributed to our progress as a Nation," his statement read. "If you want to keep up that progress, we need to stimulate scientific discovery and research, and train more young men and women for our laboratories and research centers."

Brock would find himself among many of the most outstanding scientists in those laboratories and research centers as he expanded his repertoire. Drawn to the great outdoors, his projects took a turn to ecology, particularly microbial ecology, where again he would teach himself the content and research methods. The natural world beckoned. There were ecological systems literally everywhere, even in unexpected places. Brock's attention turned to organisms that could survive in sulfur springs, and he had a passing curiosity about the creatures capable of living in hot springs. The most notable hot springs, he knew, were found in Yellowstone National Park, but Yellowstone was so . . . touristy. It had a "reputation as a heavily-visited 'amusement park,' rather than a natural area," he later wrote. His general reluctance to visit was probably the greatest career hurdle he had ever faced. Curiosity finally got the better of him, though, and he stopped there on a quick side trip in 1964. Tourist trap or not, Yellowstone boasted an array of tenacious creatures that captivated him, and the next summer Brock and his wife—also a scientist—spent two weeks sampling and culturing a variety of microscopic life-forms present in different parts of the hot springs. And on June 20, 1965, he made this note in his journal: "White Creek drainage . . . Effluent at 82°C has pink gelatinous stringy (organism?)."

Certainly, given the foot traffic through Yellowstone, others had noticed the pink material growing there. Surely people had been curious about what it was. However, perhaps those people didn't come prepared with students and equipment to analyze the chlorophyll, RNA, DNA, and protein content of what they found. The Brocks did, and what they found was that there were bacteria living—no, *thriving*—at temperatures approaching the boiling point of water. This was an extraordinary finding, given that at the time, scientific understanding of thermophiles, or heat-loving bacteria, was largely derived from laboratory analyses that never surpassed 55°C, not from field work. There was no point in creating a hotter environment, many thought, because at much higher temperatures, life couldn't possibly exist. Brock derided this approach. "The study of microbes directly in the natural habitat led to the discovery of extreme thermophiles. A reliance on . . . standard incubation temperatures of 55°C had caused investigators working up to that time to miss them." After his discovery of these extremophiles, Brock earned another National Science Foundation grant to further explore the hot spring ecosystems. Why? Because they had caught his attention, and that's the type of curiosity the National Science Foundation wanted to support.

When an honors undergraduate student approached Brock for a possible thesis project, Brock suggested something that he was personally interested in doing: cultivating the pink organisms at Yellowstone. Try as they might, they could not get the pink material to take. They did notice something else, though. While the pink microbes refused to take hold in the culture, some yellowish bacteria did thrive. Over the next few years, Brock and his students would find out more about this newly discovered bacterium, including the somewhat disturbing fact that it happily lived in household water heaters. Brock tried on a few names for size, originally calling it *Caldobacter trichogenes*, literally "filamented hot bacterium," as it formed long hair-like filaments. Before this name made it to press, the organism had become *Thermus aquaticus*, or *Taq*, for short, which made Brock wonder what the abbreviation for *Caldobacter trichogenes* would have been. Cat, perhaps? Whatever the case, by any other name it probably still would have been the source of the unlikely title of 1989's "Molecule of the Year."

14   COPYCAT

This foray into Yellowstone might seem to be an abrupt break in the otherwise continuous history of genetics, but the discovery of *Thermus aquaticus* went far beyond mere curiosity. Replicating DNA requires breaking apart an existing double-stranded sample and then fitting new nucleotides into their proper niches. This is where something like *Taq* can come in handy.

For instance, think of an incredibly short double strand of DNA with the following structure and with the lines standing in for double and triple chemical bonds:

A=T
G≡C
G≡C
T=A

If you somehow split the bonds (e.g., by heating up the DNA molecule) between the base pairs, you unravel the DNA and wind up with a single strand containing the bases AGGT and another single strand containing TCCA. Make the conditions favorable for rejoining (e.g., by cooling the DNA molecule off), and the two single strands will happily come back together as the original double strand. But when cells divide, the DNA must somehow replicate as well so that each new cell will have a full copy of the genetic blueprint. Depending on the type of cell, this might happen quite often, meaning that something has got to make exact (or very nearly exact) copies rapidly. In an average human cell, DNA is assembled at a rate of about 3,000 base pairs per minute, certainly not a casual rate by any stretch of the imagination.

The strict rules for the base-pairing allow for a surprisingly elegant solution, one hinted at by Watson and Crick, because each individual strand can serve as the template for the complete DNA molecule. If enough free nucleotides are floating around, and there is a molecular tool that can grab those and attach them to the unwound strands, the AGGT strand will have a T added to its top, then a C, then another C, then an A. Likewise, the complementary bases will be added to the TCCA strand. All that a cell needs is the machinery to unzip the DNA, plenty of available nucleotides, and a machine to click those nucleotides into place while the unwinding is going on.

The unzipping molecule, since it destroys the helix, is called helicase, following the convention of other "-ase" molecules. Helicase works like a zipper pull, trucking methodically down the DNA, unzipping as it goes. But, since the DNA will tend to naturally zip itself back up, another molecule has to come into play to hold the two strands apart. Once that is accomplished, the new nucleotides can be set into place by yet another machine: DNA polymerase. A proofreader then works its way down the newly constructed DNA molecule, pitching out mistakes, and voilà, two DNA molecules, each with one old strand and one new strand. Start the process again, and those two will become four. Those four will become eight, and so on. In fact, with this exponential progression, one could start with a relatively small sample of DNA and artificially amplify it. And, theoretically, if large enough quantities of DNA could be analyzed, scientists could sort out the actual order. A small sample of your DNA could be shown to belong to you and you alone. The problem lies in convincing the DNA to unzip and assemble quickly, and that is where Yellowstone comes in.

In the late 1960s, Brock found to his surprise that his thermophile discoveries were becoming quite the hot item. Out of the blue, scientists called him for help in acquiring samples and cultures. A storehouse of cultures and cell lines for research use, the American Type

Culture Collection in Washington, DC, has had a sample of *Thermus aquaticus* strain YT-1 since that time. A visit to the ATCC website is an almost surreal experience, with vast listings of microbes and tissues that researchers can purchase. The Brock and Freeze (Hudson Freeze was Brock's undergraduate who helped isolate *Taq*) YT-1 strain is a click, a detailed form, and $395 away from being delivered to your doorstep. A freeze-dried sample of the famous HeLa cell line, originally harvested from Henrietta Lacks's cervical cancer, can be had for a bit less (for more about Henrietta Lacks and the origin of this hardy cell line, see part V and read Rebecca Skloot's fascinating biography *The Immortal Life of Henrietta Lacks*).

The value of a culture of easily replicated human cells seems at least somewhat clear, but why would anyone pay that kind of money for a freeze-dried sample of a microscopic organism from the hot springs of Yellowstone? Moreover, why does the purchaser have to promise not to use the sample for any commercial ventures without express written consent?

Those unassuming yellow filaments of *Thermus aquaticus* turned out to hold the key to a multibillion-dollar business: rapid DNA amplification. After Kornberg's discovery of DNA polymerase, it would take another quarter of a century before its fundamental role could be exploited. By the late 1970s, scientists had figured out an ingenious method for determining the order of base pairs in a stretch of DNA. The breakthrough earned its developer, Frederick Sanger, not only a Nobel Prize, as seems to be the case for the pioneers in our understanding of genetics, but also an eponymous method: Sanger sequencing. What Sanger and his collaborators realized is that DNA is very much like an extremely long logic puzzle whose solution had to wait for the intervening discovery of a number of special molecules.

The cell is a veritable city filled with specialized workers of all kinds, working endlessly to keep the machines going, to repair damage, to alert the body of dangerous invaders. No surprise to Kornberg, scientists found that enzymes played a leading role, and it was soon discovered that a DNA strand could be snipped in particular places by specific enzymes. They also discovered a special type of molecule called a dideoxynucleotide, which behaves like a regular nucleotide in terms of its pairing (Gs pair with Cs; As pair with Ts) but turns out to be the end of the line in constructing DNA. No further nucleotides can be added, no matter how hard the DNA polymerase tries.

What Sanger and his team did was realize that they could exploit those line-ending molecules when faced with an unknown string of DNA that had a known starting point, like so:

AGGTxxxxxxxxxx

Since there are only four letters in the genetic alphabet, along with four corresponding dideoxynucleotides, they could place copies of this string in a solution containing some DNA polymerase, all four nucleotides, but only one type of sequence ender, G, for example.

The resulting string might be this long:

AGGTxxxxx

Or this long:

AGGTxxxxxxx

The endpoint indicates that the G-coded sequence-ending molecule attached there, which means that the last x must be C in each of these cases. If you follow this procedure for the three other sequence enders, you can locate the positions of all four nucleotides in the sequence:

AGGTAGTT<u>C</u>G<u>C</u>TTA

Figuring out the lengths of the resulting strings was nontrivial but essentially boiled down to a time-consuming molecular race. Sanger and his team placed the DNA strings into a special gel and applied a mild electric field. Because of the way the electric charges are situated on DNA molecules, the electric field would force the DNA to the positively charged side. The gel, however, would slow its motion down. Just as an open parachute experiences more air resistance than a skydiver despite gravity's insistence that they both fall, the longer stretches of DNA have a harder time moving through the gel. This means the shorter molecules are able to race to the end faster, just as an unparachuted skydiver will shoot past one whose chute has been deployed. With a suitably chosen gel and electric field, even strings that are only one nucleotide different in length can find themselves at measurably different places when the race is over. If this is done simultaneously for the products of each of the four possible end-points, each with its own separate "track," the exact structure can be read just like notes on a staff.

Sanger's 1977 race lasted 14 hours, and the stopping points for the different strings were determined by incorporating radioactive atoms in the original DNA mix and placing x-ray film on the final molecular race. As the radioactive tracers decayed, their image was captured on the film, revealing the location of the DNA strand with a dark stripe. Sanger's team would have found for our hypothetical string AGGTAGTTCGCTTA that the bottommost stripe showed up in the A column since the first possible endpoint after the initial 4-nucleotide segment was A. The second place stripe would have been in the G column; the third place in the T column, and on down the row. Pulling up the rear in this race, closest to the top, would have been a stripe in the A column.

Starting line

G   A   T   C
      —
            —
            —
                  —
—
                  —
            —
            —
—
      —

Finish line

In their 1977 paper announcing this amazingly simple-to-follow breakthrough, Sanger's team published images that appeared to be the score to a strange song where only the notes in the spaces of a staff were used. Two notes are never played simultaneously, and every note has the exact same duration. Notes are often repeated—four consecutive Ts, for instance—and the song can go on for thousands and thousands of notes. In the case of human DNA, it would go on for billions.

This discovery led to a huge and sudden demand for the artificial synthesis of short, known strands of nucleotides that could be used as "primers," or chain starters in DNA sequencing. Kary Mullis was a chemist doing just that at Cetus Incorporated, a biotechnology firm in California that had been founded a dozen years earlier and that was riding high on the genetic sequencing wave. However, by 1983, machines had taken over most of the work and were far more efficient than the scientists who used to be personally involved in the production. Unfortunately, creating these oligonucleotides had become almost too easy, and there were growing quantities of artificially created DNA starters but not nearly the same growth in demand.

And so late one evening in 1983, as he drove along a winding road to a weekend getaway, Mullis was lost in thought, one minute taking in the scents of springtime and another musing over the shortcomings of DNA sequencing and the very real possibility that the Cetus bubble was about to pop. Sanger's method worked beautifully—for short strands of DNA, that is. What Mullis was pondering was much bigger. He realized that to really get into the inner workings of DNA, it was going to be necessary to make huge numbers of copies of carefully selected bits and pieces of DNA. Trying to sequence the entire thing was insane, but he also felt that quickly pinpointing specific genetic strings could be valuable. And so he ruminated about the attachment of oligonucleotides, about how to trace molecular endpoints, about a number of things, when he suddenly realized that a piece of DNA—even just a minute sample—could be doubled and doubled again and doubled again, with a very quick, simple procedure: the polymerase chain reaction, or PCR.

According to Mullis's own website, "Polymerase Chain Reaction is now a word in *Merriam Webster's Collegiate Dictionary* and if you put 'PCR' into Google, you get 18,000,000 hits. If you type in 'pcr song,' you get a lovely little ditty courtesy of Bio-Rad, which will rattle around in your brain like an insane cat in your garage. Try it."

Better yet, try this recipe, which requires you to obtain a test tube, some specific chemicals, and a source of heat. The first ingredient is a sample of DNA, possibly from your skin or blood. To this you'll need to add primers—Mullis's oligonucleotides—which are short single strands of synthesized DNA that can latch on to their complementary site in the DNA sample. Two primers are required to act as bookends, as it were, on the DNA strand you're interested in copying. Then add a generous helping of A, G, T, and C nucleotides and some DNA polymerase, which will start finding homes for the stray nucleotides, one by one starting at the primer. Once your double-stranded DNA has come together, you will need to simmer the concoction at a temperature of about 95°C, which is hot enough to take your newly assembled double strand and unzip it. Now cool it to about 50°C, at which point the primer sequences will elbow their way in and attach to their prescribed locations before the separate strands can reunite. The solution then has to be heated again somewhat, and the polymerase will spring back into action. Within a matter of minutes, there will be two double strands of DNA representing the region between the specially chosen primers. Repeat the process again and there will be four. Repeat the process 30 times and there will be over a billion copies of the targeted sequence.

The entire process hinges on a few convenient facts. The first is that either strand of the DNA double helix can serve as the template to make the full molecule. The second is that the two strands are remarkably easy to separate with the addition of heat. And the last? There is an assembly molecule—a DNA polymerase—that can withstand temperatures that other DNA polymerases cannot.

15   MOLECULE OF THE YEAR

Mullis's mental breakthrough that evening saw slow realization in the lab, along with an exponential progression of controversies. His first attempts did not involve thermal cycling and failed to produce the desired results. Once cycling was employed, the chosen DNA polymerases needed to be refreshed after every cycle, as the high temperatures destroyed them. But *Thermus aquaticus* was an organism happily existing near those same high temperatures, which meant that the molecule responsible for replicating its DNA must also survive in Yellowstone's hot springs. Seeing some success in the procedure, Cetus applied for patents, which were awarded in 1987.

By then, *Taq* polymerase had been incorporated in the polymerase chain reaction that Mullis had conceived and that teams of biotechnology employees had seen to fruition. Mullis himself left Cetus in 1986, though, and not on the best of terms, so he did not share in the later success. His former colleagues described him as occasionally abusive, both verbally and physically, with an oversized personality. The feeling was mutual. In a *New York Times* interview, he was quoted as saying, "None of those vultures had anything to do with [the development of PCR]." He was given a $10,000 bonus for what, in 1986, was still an uncertain process whose potential was still far from clear. After leaving Cetus, Mullis published no further scientific papers, preferring instead the glamour of the lecture circuit.

But PCR's applications outpaced everyone's expectations, and in December 1989 the prestigious periodical *Science* declared *Taq* polymerase its first "Molecule of the Year." Interestingly, nowhere in the two-page discussion of the process of PCR and *Taq* polymerase's role in it do the authors, Ruth Levy Guyer and Daniel E. Koshland Jr., mention Mullis. By 1989, the importance of his role had become murky. How much credit should one get for mentally putting extant processes together into a new concept? Was the concept as good as the invention? Should more credit go to those scientists who took the raw concept and saw it through to an experimental design, as Mullis had not? Arguing his case, Mullis stated, "That [the realization that PCR could be done] was what I think of as the genius thing . . . in a sense I put together elements that were already there, but that's what inventors always do."

The 1993 Nobel committee apparently saw things his way, awarding him half the Nobel Prize for Chemistry that year (he shared it with Michael Smith, whose work involved "reprogramming" DNA). Even that event was not immune to controversy. While in Stockholm to receive the prize, Mullis amused himself by shining a red laser pointer on passers-by, unaware that laser sights were used on rifles and that a Swedish man had been shot with a laser-sighted rifle just the year before. He was nearly arrested. His unpopular views on the cause of AIDS (he doesn't believe HIV is the issue), astrology (he fully believes in astral planes), and alien visitation (he is convinced that a glowing raccoon, possibly an alien hologram, once said "Hello, Doctor," to him) have resulted in speculation that perhaps his position on the world stage is too prominent. One popular website even goes so far as to list him among "Four Nobel Prize Winners Who Were Clearly Insane."

The star of the PCR show, though, is not Mullis but the tiny, hardy *Taq* polymerase molecule. After its incorporation into the painstakingly determined series of solutions and cycles that must be performed to maximize the PCR yields, it became a sensation. Surely it could bask it its glory. It was a true Cinderella story. A microscopic piece of an overlooked hot springs organism suddenly becomes an indispensable part of a paradigm-shifting tool.

Taq quickly became part of a multibillion-dollar industry. Cetus sold its PCR patents in 1991 for $300 million. When that happened, the National Park Service (NPS) was understandably disappointed to be left out of the loop, given that Yellowstone was the source of the organism that yielded an essential component. Now any researchers entering Yellowstone are forbidden to use any biological samples for anything other than pure research. Should any commercial value ever emerge from their research, NPS has this to say:

> Federal law authorizes the National Park Service (NPS) to enter into benefits-sharing agreements that provide parks a reasonable share of profits when park-based research yields something of commercial value. The invention of the polymerase chain reaction (PCR) earned enormous profits for the patent owners and Diversa Corporation, which produced and sold gene testing kits that included the *Taq* enzyme. However, Yellowstone received no portion of these profits because it did not have a benefits-sharing agreement in place with Diversa.

"The Great *Taq* Rip-Off" is a mistake that NPS doesn't plan to make again, which is why if you wish to order Brock and Freeze *Taq* YT-1 strain from the American Type Culture Collection, you must first sign a form agreeing to share the results of your research, particularly if there is any potential for profit.

16   THE GENETIC BREAD MACHINE

So what exactly is so life changing about this *Taq* polymerase and the ability to quickly amplify DNA? In his original descriptions, Mullis was never particularly explicit on how this is such a boon to mankind, and indeed it appears that its potential was only realized after PCR was developed.

One of the most revolutionary things that PCR has helped bring about is the sequencing of the entire human genome in the Human Genome Project, the largest collaborative research project ever undertaken. HGP's goal, which was met in 2003, was to sequence the billions of base pairs in human DNA. The project identified over 20,000 genes, long stretches of DNA that code for specific proteins and that are the interchangeable blocks of inheritance that Mendel and Sutton had envisioned. Understanding the human genome means that scientists don't only find commonality, but they also spot when things go awry. For instance, scientists have found a particular genetic marker that indicates a marked increased risk of breast cancer. For some women who have a family history and the presence of that marker, such knowledge is agonizing. Wait for cancer to strike? Or preempt it by removing the breast tissue? Actress Angelina Jolie made headlines in 2013 when she went public with her decision to undergo a double mastectomy after learning that she carried the genetic abnormality. *Taq* polymerase was now rubbing elbows with popular culture icons, bringing visibility to the new technology bubble of genetic testing.

Another booming business, spurred in large part by child support disputes, is parentage testing. A quick Internet search will turn up DNA testing centers in your area that are willing to help people determine inexpensively who a child's parent—usually the father—is. A sample of the child's DNA is collected, along with a sample of the known parent and the alleged parent, and the presence (or absence) of several common genetic markers is checked. Half of the child's DNA will be shared with the mother, while the other half will be shared with the father. Not every single nucleotide of DNA has to be scrutinized, as the statistical rates of particular stretches of DNA in the population are known. If there's a 50-50 chance of this sequence, and a one in four chance of another sequence, then there's a one in eight chance of having both sequences. By looking at a mere handful of DNA segments, scientists can say with surprising accuracy whether two people are genetically related, a feat that is even more amazing considering that 99.9 percent of everyone's DNA is identical. Using this technique, scientists were able to fairly quickly sort out the paternity battle over the baby of the late Anna Nicole Smith, a Playboy model who died of a drug overdose in 2007. In contrast to most parentage tests, men were actually hoping to test positive for paternity to ensure a share of Smith's wealth.

More grimly, PCR has also helped identify badly degraded remains in mass graves or sites of natural disasters, events where the DNA of dozens or even thousands of people becomes comingled. DNA sequencing positively identified over half the victims of the 2001 World Trade Center attack, bringing closure to many families who never had a body to bury.

Because so little DNA is required for PCR, it has also found a role as an expert witness. Although it can't describe the events that unfolded, a dried blood spot can reveal enough to turn lives around. Since 1989, the year that *Taq* polymerase enjoyed its first taste of stardom, DNA evidence has been used to exonerate over 300 people in the United States alone, according to the Innocence Project. Evidence that was collected before PCR was even conceived has been pulled out of the storerooms for a closer, molecular look that goes beyond simply the hair texture or blood type. Where the dubious word of an informant was once begrudgingly presented as the most credible evidence, DNA, analyzed correctly, tells no lies.

The spread of television shows like *CSI* has the public believing that forensic science labs can whip up a conviction or an acquittal within an hour. The demand for quick analysis has compromised the integrity of even DNA evidence in many cases. When a Massachusetts lab technician was discovered tampering with evidence, it put into question 34,000 convictions. However, the *CSI* brand of show has also kindled great interest in the study of forensics across the country at a time when crime labs are overflowing with genetic material to analyze. In university programs, even the lower-level laboratory components now involve PCR, the "primers" purchased online and delivered via UPS. The experimental apparatus looks for all the world like a bread machine, and the instructions to the college sophomore are about as simple. "Place ingredients here. Start machine. Wait." Each heating and cooling cycle takes about three minutes, so to amass the desired billion or so copies of the selected DNA strand, an impatient student must sit through an hour and a half of thermal cycling before she can take the next step and analyze the results, a step that requires forcing the resulting strands to run the same basic race that Sanger developed decades ago. That's perhaps another hour of waiting and watching. Two and a half hours . . . practically an eternity to a busy college student.

✳ ✳ ✳   Now imagine a world where Truman had declared that all research funded by the National Science Foundation must have promise of immediate practical benefit. Imagine spending your entire life behind bars, sent there by a jury of your peers, because taxpayers found it easier to pay tens of thousands of dollars to lock you up than to pay a curious scientist to explore the organisms thriving in Yellowstone's hot springs. Imagine never knowing that the key to your innocence—and that of countless others—could be found by some college kid in less time than it takes to make a loaf of bread.

PART III

✳✳✳ Finding a Hot Spot

IMAGINE. You're 48 years old. Certainly not old, but not really young anymore. Unlike many your age, you're keenly aware that life can't go on forever. For several years you've lived with a stent—a small tubal support that fits inside your left anterior descending artery—and for several years, you've been extra cautious with your health. Today is different from the others. After a routine morning bicycle ride near the Mayo clinic, you feel the unmistakable crushing sensation in your arm. You know you need help, and you need it now. You turn to the person nearest you, who fortunately happens to be a doctor. Within seconds he is able to pull up your electronic medical records on his tablet, thanks to a recent Mayo initiative and the ever present WiFi in the area. An aspirin arrives first, then a dose of blood thinner. The paramedics arrive and begin an EKG on you, and the doctor compares the output tape with the image from your medical history on his iPad. His suspicion is confirmed. The artery is clogged again. Within minutes, a team meets you at the hospital with orders to remove the clot, which you later find out was blocking 90 percent of the artery. Had you visited the ER first, the doctor tells you, the initial test would have come up negative, adding hours to your diagnosis and resulting in major heart damage or, more likely, your death. Instead, just three days later, you are out and about again, meeting the doctor whose instant access to your medical records saved your life. As you hug him and the others who helped you, the farthest thing from your mind is the possibility that tiny black holes swarmed the early universe and that they might be evaporating as you express your gratitude. However, the failed search for these exotic objects 35 years prior is one reason you are alive today.

17   A UNIVERSAL "HOT SPOT"

Black hole. It's the modern term for what was originally called a gravitationally completely collapsed object. While the old description was more accurate, the newer expression practically exudes mystery. The public never clamored to better understand gravitationally completely collapsed objects, but one of the first things people ask astronomers they unexpectedly meet is something about black holes. Hollywood has devoted entire movies to these objects, often depicting them as ways to travel through space, time, or both, thus increasing the public excitement and, sadly, the misconception. So intriguing are black holes that magazines sell more copies when artists' illustrations of black holes, phenomena we've never actually seen with our own eyes, are featured on the cover.

Even within the scientific community itself, black holes seem daring and exotic. The computer coding and mathematics skills required to study these bizarre objects are held by only a small subset of trained astronomers. It should be no surprise, then, that the opportunity to make an amazing discovery about black holes would drive some of the most mathematically inclined people in the field.

Black holes really are the new black.

But . . . they're not really all that new. Their existence was first suggested by parson John Michell in 1784, so the idea percolated for quite some time before anyone could do anything about it. Largely a quaint mathematical curiosity, black holes had to wait for Albert Einstein's general theory of relativity in 1915 to finally give scientists suitable theoretical tools to handle them (and even those tools fall short). In 1939, J. Robert Oppenheimer and Hartland Snyder made some relativity-based calculations to determine what would happen to the core of a massive dying star. What they found was that for a suitably massive star, the core would keep shrinking forever. "The star thus tends to close itself off from any communication with a distant observer; only its gravitational field persists," they wrote. It was the first hint that, instead of an abstract concept, a black hole could very well be something real and something common. Oppenheimer never really followed up on this exciting prospect, as he was known to flit from one captivating project to another, rarely settling down on any single topic. Then, in the early 1940s, he was asked to forfeit his astrophysical pursuits to become involved with the Manhattan Project. For this, he would later become known as the father of the atomic bomb.

Oppenheimer would die four years before the first black holes were detected. In 1971, the x-ray satellite *Uhuru* caught a glimpse of a strong x-ray beam coming from the constellation Cygnus, a tell-tale sign of a gravitationally completely collapsed object. Dubbed Cygnus X-1, this source is actually a two-member system consisting of a black hole containing 15 times the mass of our Sun and a colossal star with a temperature of 30,000°C and the light output of 385,000 Suns. The black hole is ravenously consuming the outermost gas of its companion star. As the matter from the supergiant star spirals toward the black hole's unforgiving maw, the fast-moving gas releases the unfathomably high energies of x-rays before slipping forever out of sight. What we witness is a region of intense x-rays from the superhot material. A universal hot spot, if you will.

This type of black hole is practically cliché nowadays. The natural remnants of the deaths of massive stars theoretically should pepper the entire universe, and in 1974, an article appeared in the *Astrophysical Journal* entitled, "2U 1700-37: Another Black Hole." Another one? Already? Apparently the honeymoon was over.

Around the same time that star-massed black holes were becoming well known, astronomers began speculating that an entirely different type of black hole might exist. These objects were dubbed supermassive black holes, and they made their homes in the cores of large galaxies. In fact, it was suggested that such black holes were the driving force behind the violent activity seen in the brightest young galaxies billions of light-years away. The debate was quite lively for years, and the first strong evidence of a supermassive black hole in the center of a galaxy appeared with the 1993 work of Naomasa Nakai and colleagues. The case was seemingly settled when Hubble Space Telescope data revealed signatures of black holes weighing millions, sometimes even billions, of times the mass of the Sun nestled deep within the hearts of closer galaxies. The data were so compelling, in fact, that in 2000, black hole researcher Hans-Walter Rix stated, "It is quite clear that the nearby universe is full of black holes in galaxies. Using Hubble we now routinely find black holes in perfectly typical, normal, boring galaxy centers. Indeed it seems likely, that Nature can't make a big galaxy without a black hole in the center."

Now over a decade later, supermassive black holes—along with their stellar-remnant counterparts—are commonplace, so much so that even the most basic astronomy textbook would be shunned for failing to mention both types of black holes. But there is another type of black hole whose story hasn't been so widely told. Since the first discovery of a stellar black hole, theorists and observers have been trying to track this more elusive prey: the oldest and smallest black holes whose story goes back to the beginning of the universe.

18 CLASSIC BLACK

Go outside and take a good look at the night sky. The first thing that leaps out to the casual observer is that the darkness is punctuated by small dots—stars—where matter is concentrated. These concentrations have gaping chasms of near nothingness (at least, as far as our eyes can detect) between them. In fact, if the Earth-Sun distance were scaled down to a couple of centimeters (an inch) and both the Sun and Earth were reduced to specks, the next nearest speck (Proxima Centauri) would be 6.4 kilometers (four miles) away. With the aid of a telescope we can see island universes called galaxies, again with vast distances separating these relatively puny and dense clumps of matter. Galaxies cluster together on even grander scales, concentrated filaments and blobs of mass among unimaginably great voids of space.

These are fairly basic observations of the current universe. What is harder to see is what these grand concentrations of matter are doing. Edwin Hubble found in the 1920s that, by all appearances, galaxies are rushing away from each other and from our own Milky Way galaxy. This is not a motion that we can watch unfold, as the scales are so great that none of it has changed noticeably over the entire history of humankind. Instead, we take advantage of a simple way to gauge the speed of the objects in the universe, something we've been doing since the 1840s, when Christian Doppler realized that the changing pitch of trumpet players on a moving train told us about something more than music.

Just as the note from a horn played on a passing train seems to shift from a higher pitch (corresponding to shorter sound waves) to a lower pitch (corresponding to longer sound waves) when the train passes, light from moving objects in the universe will display shifts that depend on their motion relative to us. Doppler knew this should be the case, but in the 1840s, technology was not up to the task of discerning this wavelength shift in light.

Fast forward to Hubble, who was lucky enough to live in a time when the subtle shifts to longer or shorter wavelengths of light could be detected and measured. When looking at 46 galaxies, he found that the light from all but one of them showed a wavelength shift to longer, redder waves. In other words, the galaxies' spectra were redshifted. And just as the lower pitch of an ambulance siren means it's moving away from you, the redder wavelengths of the galaxies seemed to indicate that they are moving away from the Milky Way.

As if that weren't weird enough, Hubble also discovered that the more distant a galaxy was, the larger its redshift was, which only made the situation more bizarre. Why should the Milky Way appear to have such a central location? And how do those distant galaxies know how far they are away from us so that they know to adjust their speeds accordingly?

Over the next few decades, cosmologists took a deeper look at the redshift-distance relationship and came to the conclusion that galaxies are not rushing away from us like speeding trains but that space between the galaxies is expanding. It is as though galaxies are riding along on a sheet that is being stretched. When the sheet doubles in size, the distance between any two galaxies doubles in size, although the galaxies themselves do not grow. In this manner, the more distant a galaxy is, the "faster" it will appear to be leaving the scene. More important, it turns out that Hubble's observation would be true no matter which galaxy you happen to be living in. If the distance between my galaxy and yours doubles, then so does the distance between your galaxy and mine. Thus, our special central location is just an illusion.

It didn't take much of a conceptual leap to imagine that this stretching, expanding, matter-and-energy-filled universe that we see today must have started as a very small, energy-filled, dense universe quite a long time ago. Qualitatively, the story of the universe is easy to follow, but mathematically it becomes much more involved, particularly the instant after its "birth." Throughout the 1940s and 1950s, theorists worked feverishly to figure out how a vast universe could have been born and grown to its current state.

It wasn't trivial, but modeling the birth and expansion of a vast universe was not completely impossible. A tiny hatching cosmic egg that gave rise to an enormous cosmos had already been suggested by Belgian priest and cosmologist Georges Lemaître in the 1930s. By the late 1940s, the team of Ralph Alpher, Hans Bethe, and George Gamow—a trio whose surnames sounded like the Greek letters alpha, beta, and gamma ($\alpha\beta\gamma$)—described how a once-dense, expanding universe even made sense on an elemental level. With its origins in a superdense, superheated neutron fluid, our expanding universe would have naturally created the elements we see around us in the proportions we see: heavy on the easy-to-build elements hydrogen and helium and sparing on everything else on the periodic chart.

Not everyone was convinced. Astronomer Fred Hoyle explained the ratios of the elements by having them be cooked them up in the interiors of stars. The universe he envisioned was one where matter was continually created, pushing the universe apart to give us the redshifts. Hoyle's universe had always been and would always be this way, eternally self-generating and infinitely big and old. He would have none of this nonsensical cosmic egg, stating that the trio's "big bang," as he pejoratively called it, was "an irrational process that cannot be described in scientific terms . . . [or] challenged by an appeal to observation."

But there *was* an appeal to observation. Buried inside an article that was written to correct a paper on universal evolution by Gamow, astronomers Alpher and Robert Herman reported that a hot, dense initial universe should have left a thermal fingerprint everywhere. They estimated that there should be light throughout the universe with the temperature profile of an object –268°C, which is equivalent to –451°F and a mere 5 degrees Kelvin above absolute zero. For a decade, they begged the radio telescope users of the astronomical community to go out and look for this temperature profile, to no avail. To be fair, it wasn't completely the fault of astronomers. Radio astronomy was a newborn science, and many said it simply couldn't be done with the technology of the time.

Then in the 1960s, the beautiful discovery was made, not because of the urgings of Alpher and Herman but because Bell Laboratory wanted to improve intercontinental communications. To do this, they constructed a 15-meter-long, 6-meter-wide (50-foot-long, 20-foot-wide) "horn reflector," something that looks a bit like a behemoth antique gramophone. The idea was that 30-meter-wide (100-foot-wide) Mylar balloons would be sent into orbit. Radio waves could then be bounced off them and sent halfway around the world. Project Echo, as it was known, was a major step toward modern-day satellite communications, impressive enough to earn a spot on U.S. postage stamps. Strangely enough, it was also a major step toward our understanding of the universe. During the year that radio astronomers Arno Penzias and Robert Wilson tried to calibrate the antenna so that it could detect the faint signals from the balloon echoes, they found something interesting. No matter what the season or time of day or direction of the antenna, there was an inexplicable excess signal. Arno Penzias and Robert Wilson reported their findings in a somewhat apologetic paper in 1965, which dryly outlined the various steps they had taken to get rid of the mysterious "excess temperature," which amounted to about 3.5 degrees Kelvin.

Without looking for it, they had found what Alpher and Herman had predicted two decades earlier: the cosmic microwave background (CMB). It was the leftover energy from the early universe, energy that didn't become tied up as matter but instead stretched with the expanding universe. Filling the entire universe throughout nearly all of time, it was seen as the smoking gun for the big bang theory, a prediction that Hoyle's steady-state universe could not neatly account for.

While the CMB seemed to answer one fundamental question—whether the universe had a "beginning"—it gave rise to many more. Why was the background radiation so astonishingly uniform in every direction? If it truly represented the energy that did not condense into matter, it should show variations in intensity just as the matter in the universe showed variations in density. And yet what Penzias and Wilson found seemed perfectly, perplexingly smooth. How could we then model a universe that was clearly clumpy but overall extremely uniform?

An entire subfield devoted to the necessary primordial irregularities and perturbations popped into existence like a tiny intellectual universe in the 1960s, and it has been expanding ever since. Terms like "quantum fluctuations" and "inflationary universe" were coined as theorists attempted to address the issue. What they ultimately realized is that ordinary, seemingly uneventful empty space is actually an extremely dynamic place. Particles can and do pop into and out of existence on unimaginably short timescales. Thus the early universe would have been a complete mess of particles, antiparticles, virtual particles, and photons interacting, disappearing, and appearing. On top of that, it appears to have had an incredible amount of something called dark matter, something that interacts gravitationally with everything else. This mysteriously invisible matter worked behind the scenes, setting the stage for a clumpy universe filled with concentrations of stuff amid great voids.

Consider for a moment the image some have made to help visualize the weblike maze of virtual global connections. The early universe would have looked strangely like the Internet at this stage. The Internet is a perpetual haze of content creation and destruction (think "Error 404," which you see when you click a link for a page that no longer exists). Yet sometimes, a piece of content that is not fundamentally different from other pieces of content enjoys more "traffic." Similarly, random variations in particles made of normal matter and the elusive dark matter would cause a slightly higher density here, a slightly lower density there, until voilà: the scaffolding of future galaxies was created. A period of rapid inflation would then smooth most of space out, resulting in the largely featureless CMB.

Although many cosmologists were content to call it a day at that point, a young and recently wheelchair-bound Stephen Hawking saw the implications more deeply. With a mind that was constantly on the move, he soon realized that the consequences of these fluctuations went beyond providing the framework for the material universe. In fact, large enough variations in a small universe could lead not just to the matter distribution we observe in galaxies (and now in dark matter) but also to high concentrations of matter in some places and low concentrations in others. These highly concentrated places would be the cosmic equivalents of Internet memes (look up "Grumpy Cat"), a vanishingly small piece of the Web that bafflingly attracts an incredible amount of activity. It is in these smaller-than-atom regions, Hawking postulated, that so much mass might gather that even the rapid expansion of the universe would be insufficient to pull the matter apart. This matter would be so concentrated, in fact, that even light could not escape.

The universe's first black holes.

These black holes could have everything from tiny masses of ten micrograms or so up to the more respectable planet-like masses of 100 billion-trillion kilograms. Even more interesting, they might be swarming all over the universe, the smallest ones capable of passing unperturbed through a universe full of solid lead. Tiny in size (black holes on the high-mass end of this spectrum would span less than a tenth of a millimeter; the smallest ones would be less than an atomic radius in size, or a millionth of a billionth of a meter) and virtually undetectable, primordial black holes have been implicated in everything from the missing dark matter in the universal census to terrestrial ball lightning to sporadic solar oscillations.

With a single relatively brief paper in the *Monthly Notices of the Royal Astronomical Society*, Hawking conceived of something even more intriguing than a regular, run-of-the-mill black hole. He came up with the oldest fossil remnants of the universe.

Then he went a step further. The same variations and uncertainties that brought them (and the structure of the material universe) into existence could also apparently be their undoing.

19   A TUNNEL TO OBLIVION

Picture a ball at the bottom of a deep ditch. A child comes along and kicks the ball, but, if the kick doesn't give the ball enough energy, it will still be confined to the ditch. In our everyday, macroscopic world, nothing would be weirder than to see the ball appear outside the ditch, especially if nobody had kicked it.

In the microscopic world of subatomic particles, however, the occasional ball does find itself on the banks of the ditch or otherwise beyond a barrier that it had no obvious way of getting through. This phenomenon is known as "tunneling." It sounds bizarre and counterintuitive, but quantum tunneling turns out to be a somewhat predictable thing that we have gleefully exploited in the development of semiconductors. But tunneling out of a black hole? It had not really occurred to black hole theorists to bother incorporating something acting on such a minuscule scale, both spatially and materially, in their computations. It would be akin to accounting for the possibility that the Colorado River could suddenly appear on the rim of the Grand Canyon. Sure, it is physically allowed according to the rules of quantum mechanics, but it's statistically pointless. Hawking himself opened his 1974 paper explaining why it's safe to ignore these so-called quantum gravitational effects for most black holes most of the time.

Science thrives on exceptions, though. Most black hole theorists were concerning themselves with the collapsed cores of massive stars, black holes whose masses are measured as multiples of the Sun's mass (the Sun's mass is two billion billion trillion kilograms). The primordial black holes, however, theoretically could have masses like everyday objects and gullets smaller than a proton. In these environments, tunneling particles could slowly, surely make a world of difference.

Hawking computed that black holes with sufficiently small masses would also have sufficiently small gravitational territories. So small, in fact, that the same quantum fluctuations that allowed particles to pop into and out of existence in the early universe could also allow particles to find themselves outside the black hole that they once were part of. And, thus, the black hole becomes ever so slightly less massive.

A less massive black hole—even just an electron less—is technically a *smaller* black hole, just as the Grand Canyon that has had a shovel of dirt tossed into it is technically a shallower, more easily escaped Grand Canyon. This means that it's even more likely that a particle can find itself outside the grips of the black hole. And then the black hole is even less massive. And so it goes.

This sort of process doesn't happen overnight. Even a tiny black hole with a mass of a few billion tons and the width of a proton could endure for ten billion years or so (the more massive ones would last what might as well be forever). But those last moments would be epic. As the black hole lost mass, the more quickly it would shed its own mass through "evaporation," a word that is quite possibly the poorest description in all of science. As Hawking points out, in the last tenth of a second of its existence, one of these dying primordial black holes would emit as much energy as a million megaton hydrogen bombs. While this is piddling compared with your basic supernova explosion, which releases ten trillion trillion times more energy, it should still be noticeable. Furthermore, since the universe is in the neighborhood of ten billion years old (13.8 billion, to be more precise), these primordial black holes should be evaporating all around the universe as you read this.

20   CHASING WILD GEESE

If the story so far seems comfortably removed from anything that could possibly affect your life (unless one of those primordial black holes shot through your backyard), it's because it is.
At least it was until the moment that famed astronomer Martin Rees,
who would later be knighted as Baron Rees of Ludlow, considered in
a 1977 paper the methods by which we might observe these things.
He pointed out that, although the evaporating black holes would
produce insane amounts of high-energy light called gamma rays in
their last moment of existence, the most practical way of observing
them might counterintuitively be in the low-energy domain of radio
waves. These radio waves would be produced not by the black hole
evaporation itself but by the interaction of the shock wave created as
the energy and particles released in the grand finale crashed through
the low-density material and their associated magnetic fields in the
surrounding neighborhood. A "crude, non-directional antenna," he
figured, could see such events from as far as 30,000 light-years away.
"And an Arecibo-type telescope could detect such pulses—each triggered by a single entity of subnuclear size—from as far away as the
Andromeda galaxy," which is over two million light-years distant.

A more exotic alternative recently suggested is that a radio signal
might be produced by the encroachment of the black hole into a
fourth spatial dimension. This in itself is sufficient proof that black
hole theorists concern themselves with things far beyond the realm
of normal daily existence. Further proof is found in this statement
from a 2009 paper discussing this intriguing transdimensional prop-

erty: "The change in topology of the black hole is easy to understand." Or so Patterson and his co-authors claim.

In 1977, radio astronomers Peter Shaver and Ron Ekers, a Canadian and an Australian, respectively, who were both researchers at the University of Groningen in the Netherlands, were up to Rees's challenge of finding the tell-tale signs of an exploding black hole. Ekers had already been involved with Rees in a radio wavelength search for the supermassive black holes in the hearts of galaxies, and he was intrigued by the chance to make a game-changing discovery. Meanwhile, Shaver was an expert on the interstellar medium, the low-density material between the stars, and knew how that material could alter the signal from exploding mini black holes. Together they decided to be on the lookout for the dying screams of evaporating black holes.

The good news was that Rees had estimated that there might be as many as 100 billion trillion diminutive black holes in our galaxy, so even a modest survey had a good fighting chance of seeing something, particularly if it concentrated on a part of the Milky Way known to have an abundance of matter. The bad news was that Rees had also conceded that there might be none. These were, after all, only theoretical objects that might be everywhere or might be nowhere. Worse still was that the specific radio blip predicted might not even be produced by the theoretical deaths of these theoretical objects. It was a long shot to the third power.

Luckily for them, the signal that might or might not be produced in a death that might or might not happen to an object that might or might not exist would absolutely have very specific characteristics. In its death throes, a dying black hole should emit a great range of wavelengths, but with an infinitesimally short duration: only a single wavelength cycle, or about a billionth of a second. Shaver and Ekers knew early that they would need something that could detect a shorter pulse than had ever been picked up by radio astronomers. On top of that, although the original burst of radio waves would last less than an instant, the lower frequency part of the signal would be delayed as it traveled through the interstellar medium. In a sense, the longer radio wavelengths get slowed down shaking hands with all the intervening electrons, while the shorter wavelengths manage to pass through relatively unimpeded. The same sort of process is responsible for rainbows. With visible light, the different colors (wavelengths) interact differently with the water droplets, causing one color (in this case the shorter, bluer wavelengths) to bend more as the light passes through. The amount of refraction, or delay in this case, depends on both the wavelength of light and the material it's traveling through.

The upshot is that instead of getting a sharp, multi-frequency signal lasting only a billionth of a second, astronomers would observe a longer-duration signal beginning with the higher frequency waves and ending with the lower frequency ones—a down chirp, in radio terms. If you have ever heard a hawk's piercing cry that starts at the highest note and trails to lower pitches, you have heard the audible (and greatly protracted) equivalent of the radio signal of a dying black hole.

To get an idea of how this plays out on the receiving end of a black hole's death, imagine a musical chord in which several notes (frequencies) are played simultaneously. A single note has its own fairly clean waveform, which can be represented by a sine wave for a pure note. But the combined waveform from all the different frequencies creates a visually chaotic mess of wiggles. As it happens, there is a fairly straightforward mathematical way to extract the individual notes. Called a Fourier transform, after French mathematician Joseph Fourier, it's a mathematical black box that allows any signal to be represented as a combination of clean sine waves of different frequencies. A fast Fourier transform (FFT) is a mathematical trick designed to perform the task more quickly, hence its name. In this way, a chord can be broken down into its component notes, along with the volume level of each individual component. If there is some change to the chord over time, this information, too, can be extracted.

What Ekers and Shaver faced was the cosmic equivalent of an

entire piano keyboard of frequencies that, although played at precisely the same time, did not arrive at our ears at precisely the same time. Instead of the 88-key strike, we hear a quick glissando down the notes, or a down chirp. To top it off, humans create quite a bit of radio waves on their own in the frequency range that was being explored. So, in addition to the full-keyboard piano strike that they were looking for, they had random musicians all over the place. Consequently, what they needed was someone with both the science background to fully understand the problem and the engineering skills to devise something that could pull useful information out about the various frequencies and their arrival times *and* do this amid a teeming background of man-made signals.

They needed someone like John O'Sullivan, an engineer who was coincidentally also an Australian working in radio astronomy in the Netherlands.

It was the type of collaboration that is second nature to scientists and engineers. Although his training was in mathematics and engineering, O'Sullivan had been part of the radio astronomy community since his undergraduate years at Sydney University in the 1960s. To him, the field of radio astronomy struck the perfect balance between scientific curiosity and the practical issues involved in exploring that curiosity. The chance to work on a project with such profound cosmological implications, even a project with the odds stacked so heavily against success, was not to be missed.

Their search began in the northeastern Dutch province of Drenthe, home of the Dwingeloo 25-meter (80-foot) radio telescope. Looking a bit like a mobile home straddled awkwardly by the skeleton of a giant satellite dish, this large radio telescope was conceived in 1944 by famous Dutch astronomer Jan Oort, who wanted to observe a specific radio wavelength—the 21-centimeter line—associated with cold, neutral hydrogen between stars. It took 12 years for the ambitious project to reach fruition, but when it was built, it was a state-of-the-art instrument that could explore in great detail the element that makes up most of the material universe.

By 1977, though, the Dwingeloo telescope was becoming almost passé. The more modern Westerbork Radio Telescope made use of multiple, widely spaced dishes that acted as a single large dish. But the lone Dwingeloo dish was a good starting point for the full-time radio astronomers to work this project into their usual observing routine.

O'Sullivan's first engineering project for the search was simple: make the existing equipment do something it was never designed for, a charge that is practically every engineer's raison d'être. In this case, O'Sullivan was given the daunting task to create something that could detect a very short pulse of many wavelengths, something never done before. At first the team started small, trying to detect only two of the myriad frequencies that were expected to emanate from a dying black hole. Their team grew, and so did the number of "channels" or frequencies they were trying to monitor. Their list of observing targets included many familiar objects but also a number of places where black holes might be expected to gather. With the unprecedented ability to observe fleeting radio pulses, the team looked at everything from Jupiter to the outskirts of our galaxy to distant quasars.

The design and construction of the hardware for the original project took weeks, but it took only a couple of days to carry out the observations. Although a complete shot in the dark, it was all part of the ongoing research funded through the Netherlands Foundation for Radio Astronomy. It had not a hint of practical application and even less potential profitability. As for answering their scientific question, the researchers detected precisely zero exploding black holes.

Their null result was reported in a benignly titled 1978 paper: "Limits on Cosmic Radio Bursts with Microsecond Time Scales." Then in 1980, the 300-meter-wide (1,000-foot-wide) Arecibo radio telescope, a skateboarder's dream nestled in the caldera of a long extinct volcano, seemed to spot some tantalizingly short pulses originating in another galaxy. At over 53 million light-years away, the galaxy dubbed Messier 87 rekindled the interest in exploding black holes. But 53 million light years of intervening material would distort a dying black hole's signal enormously. The device designed for Dwingeloo simply wouldn't work on a pulse that was that badly smeared out, and so O'Sullivan went back to the drawing board.

A creative but unwieldy solution was dreamed up by Tim Hankins, who developed something called an opto-acoustic Fourier transform machine. This involved a convoluted setup whereby the radio signals were translated into vibrations within part of the instrument. A laser beam shining into the now-vibrating medium would then help trace out the various frequency components onto, of all things, 35 millimeter movie reels. The team had worked out just how the signal should look, and they spent untold hours poring over miles of film searching visually for the tell-tale diagonal.

This was the last straw for O'Sullivan. "There has to be a better way," he told himself, his colleagues, and everyone else who would listen. The very next week he began work on a digital solution to the problem.

And still, no exploding black holes had been found.

21   GOING WIRELESS

In 1983, Bob Frater, then Australia's Commonwealth Scientific and Industrial Research Organization (CSIRO) Institute director, all but begged O'Sullivan to return to Australia to begin looking at possible commercial applications of the work he had done in radio astronomy. The same year saw the invention of the cell phone. A piece of Motorola hardware lovingly termed "the brick," it had a whopping one hour of talk time and could stay in stand-by mode for nearly eight hours. It was a time when home computers were clunky boxes that spoke to each other through cables or through the technological miracle of the modem, where staticky signals passed through corded telephone handsets. The future we currently inhabit was beyond most people's imagination, but Frater knew that extracting useful information from distorted wire-free signals could be extremely valuable in fields like medical imaging, geophysics, and telecommunications (not to mention the military applications). Manipulating and interpreting high-speed signals could only prove to be a boon.

While at CSIRO, O'Sullivan led a project with Austek Microsystems that, in 1987, resulted in a single chip that could quickly perform fast Fourier transforms: the A41102. It was to the cumbersome hardware at the Dwingeloo and Westerbork radio observatories what the laptop is to the computers of the 1980s. Designed solely to clean up frequency information from an input mishmash of waves, this minute block containing 160,000 specially arranged transistors outdid the supercomputers of the time at its task.

Computer technology progressed in leaps and bounds, but networks were perpetually tethered to each other with physical connections. Tons of cables were manually dragged through the ceilings of research labs, universities, businesses, and eventually houses. The most intricate and time-consuming gymnastics had to be performed during any kind of renovation or relocation. At this time, taking the computer to a working lunch would have required plenty of wire. Or better yet, lunch delivery.

It wasn't as though information couldn't be transmitted from computer to computer via radio waves. It was just a matter of speed, as the signal inevitably gets tangled up with itself on its way from one computer to another. The only solution at the time was to send out information slowly so that each piece could be suitably translated before the next piece arrived. Such engineering issues have always plagued communications. After all, signal degradation was the underlying reason it took 16 hours get a 98-word message through the first trans-Atlantic cable in 1858.

The problem with wireless computer networking, though, is not an ocean's width of physical cables, as in the case of the trans-Atlantic lines, but walls and furniture and electrical wires that all have the effect of causing the signal to echo from a huge number of barriers. If you shout at a flat, distant mesa, someone next to you hears your initial shout and the delayed echo. If you shout in an irregular canyon, your voice will hit the person next to you numerous times at different volumes. And if you attempt to sing "Twinkle, Twinkle Little Star" in an irregular canyon, you might as well be singing rounds all by yourself, as one verse will make it back to your friend as you're singing the second. On top of causing the radio waves to "echo" around the room, the obstacles treat different wavelengths differently, just as the material between stars treated low-frequency waves and high-frequency waves differently.

But this is a problem that O'Sullivan had already largely figured out. Radio waves don't travel cleanly around a room or a building any better than they do through the stuff between the stars. "Looking at the problem in the frequency domain was similar to the thinking which we applied many years before to the dispersed pulse detection," O'Sullivan explained. Instead of sending information from computer to computer as a single stream of data, a stream that would inevitably turn into a radio-wave rendition of "Twinkle, Twinkle Little Star" in a canyon, he saw that the information could be sent out simultaneously as multiple streams at different frequencies, each one containing only a fraction of the whole message. The message could then be reassembled quickly at the receiving end.

To appreciate this, imagine that you have tickets to your local orchestra's performance of "Queen's Greatest Hits," a fundraiser to help repair the concert hall following a natural disaster. Unfortunately, the damage from the disaster left the hall with the worst acoustics in the entire history of music. High notes echo strangely, and low notes, for whatever reason, take longer to reach your seat. The orchestra's playing is flawless. Your enjoyment of it . . . not so much. The conductor, though, is also a brilliant and hyperkinetic engineer. Her solution to the resulting cacophony is to digitize the music so that all the B-flats are transmitted on one radio station, all the high Cs on another, all the F-sharps on another, and so on. Each concertgoer is given a special set of headphones that isn't tuned to any particular radio station. Instead, the headphones simultaneously pick up every one of them. On top of that, they have a device that can reconstruct the entire concert the way the orchestra plays it. Even though no single channel is carrying the entire concert, the various signals can be reassembled to remove the distortions created by the environment. Now take this analogy and mix together the talents of physicists, engineers, radio astronomers, mathematicians, and computer scientists, add a handful of time, and you have something like the real story.

O'Sullivan, who by now was focusing on more terrestrial applications of his work, formed a team with Terry Percival, one of many who saw great potential in rapidly sending and receiving information wirelessly. Percival was a communications engineer who, like O'Sullivan, had made his entry into engineering through radio astronomy, a field that has proven to be a great training ground. His career path had taken him far afield to the iconic Karl G. Jansky Very

Large Array in New Mexico, an enormous Y-shaped collection of radio telescopes later made famous in the 1997 movie *Contact*. He then found himself in the world of commercial satellite telecommunications, briefly working with the timing standards from GPS satellites, whose history also is tied to curiosity-driven experiments.

Back at CSIRO, Percival and O'Sullivan joined forces with computer programmer John Deane, scientific jack-of-all-trades Diet Ostry, and Graham Daniels, a circuit designer whose interests ranged from microprocessors to system security. The group began with the idea of the A41102 chip and began developing a full wireless local area network system, a.k.a. WLAN, an acronym you might have seen when attempting to troubleshoot the connectivity issues in your laptop or tablet. By 1992, they applied for a U.S. patent, and in 1996, the team was awarded patent 5487069 for the first WLAN. In 1997, a generation after the first failed search for exploding black holes and a full decade after the first fast Fourier transform chip, the device and system described in their patent became the basis for what is cryptically known as IEEE 802.11, the industry wireless network standard set by the Institute for Electrical and Electronics Engineers.

22   WHERE CREDIT IS DUE

At this point, the made-for-TV movie on the development of WiFi would show the scientists and engineers popping champagne amid confetti and closing credits. Life at the cusp of a multibillion-dollar industry rarely wraps up so neatly. CSIRO wasn't the only place on Earth exploring wireless capabilities. The surge in portable devices in the late 1980s and early 1990s had R&D programs at several companies working at a frenzied pace, and by the time O'Sullivan's team applied for their patent, the handwriting was on the wall. The concepts used to transmit and receive useful signals were based on physics, not magic, so the same ideas were cropping up across the globe. Fast Fourier transforms might have been impressive and esoteric mathematics in the early 1970s, but by the 1990s, they were a standard topic in undergraduate engineering courses. As for multipath interference—the smearing out of the signals as they bounced around a room—this, too, was well known in the industry.

CSIRO would later file for and obtain patents in 19 countries, not including Russia, China, or India or those in Latin America (they would later lament not filing for patents in China and India). But CSIRO was not a manufacturing company. While their mission included the development of new technologies, it would require the creation of a dedicated company to put together a marketable device. This company was Radiata, formed in 1997 by David Skellern and Neil Weste of Macquarie University in northern Sydney. O'Sullivan was hired as vice president for Systems Engineering, and it was actually this company that would create and demonstrate the first actual chip set based on the IEEE 802.11 standards. It was lightning fast compared with the equipment in most people's homes at the time. In the olden days of phones physically cradled on the computers, users could expect data to be transferred around 300 bits per second, or 300 baud. When computer memory was measured in kilobytes (thousands of data units known as bytes), as opposed to the gigabyte memories of current household laptops (giga = billion), this seemed pretty cutting edge. Wired modems that transferred data a hundred times faster in the 1990s seemed practically miraculous. But this wireless technology could receive and transmit data at a rate of 54 million bits per second. This was astonishingly fast!

Their demonstration floored the tech world. Having forked over four million Australian dollars to help start up Radiata, Cisco Systems went whole hog in November 2000 and bought out the then 54-employee-strong company for the impressive sum of 567 million Australian dollars. This was bad timing, as everyone was by then in the race to make computer networks wireless. Although CSIRO allowed Cisco to use the WLAN technology (as they would have for any company that sought licensing from them), CSIRO still held the patent. As such, they were pretty insistent that they receive appropriate royalties from any companies using technology underpinned by the IEEE 802.11 standards, which is to say pretty much everyone.

At first, CSIRO attempted to negotiate for the royalties they felt they were entitled to. It became quickly evident, though, that the WLAN horses had long since left the barn. CSIRO filed an uncharacteristic lawsuit in 2005 against Buffalo Technology, a company that most people have probably successfully ignored. CSIRO's victory opened up a rapid and ugly set of disputes with every company that incorporated WiFi in their devices. Big names like Dell, Microsoft, and Hewlett Packard went to battle, and, by 2009, 14 major companies were being sued by CSIRO for copyright infringement. CSIRO was vilified as a patent troll, attempting to profit off a patent that was more a "state-of-the-technology" report than a bona fide invention. Companies would argue timelines, prior art (whether the information in the patent was already well known in the industry), and whether CSIRO could legally sue, given that they were not even a competing manufacturer of the products in question.

Hewlett Packard was the first to offer a settlement in a move that spelled victory for CSIRO. The first round of companies shelled out over $200 million in royalties. CSIRO didn't stop there, though. Computer manufacturers Acer, Lenovo, and Sony, along with cell phone giants Verizon, AT&T, and T-Mobile, were also marketing WiFi-equipped devices. Settlements with these companies have also topped $200 million, yielding well over $400 million for CSIRO by mid-2012. That's not a bad sum, considering CSIRO's annual science budget runs about $1.4 billion, most of which is government supported. Research done there, even on something as theoretically obscure as exploding black holes, is in a very real sense paying for itself. As the number of WiFi-enabled devices approaches the population of planet Earth, CSIRO stands to gain even more.

23 ANATOMY OF A SUCCESSFUL FAILURE

Scientifically, this has been a tale about a failure. Whether it's a failure in just practice or in both theory *and* practice remains to be seen, but at this point no exploding black holes have ever been unambiguously observed. John O'Sullivan has returned to CSIRO and is now working on an ambitious project: the Australian Square Kilometer Array Pathfinder (ASKAP), a pricey and, as its name suggests, square-kilometer (0.4 square-mile) array of 36 radio telescopes in Western Australia. A man who has been involved in pure research, applied research, start-up companies, and established corporations, O'Sullivan sees human activities as a sort of ecosystem. "If we cut any one of those levels back, we threaten the existence of the whole ecosystem. Not all pure research activities will lead to applied research, and many applied research efforts will lead nowhere," he argues.

Echoing this sentiment is Dr. F. Fleming Crim, the assistant director of the National Science Foundation's Mathematical and Physical Sciences branch, who used a horse race as an analogy for fundamental science research projects. "If you want your horse to win," he explained, "you have to start a lot of horses." In fundamental research, the horses that don't win are comparatively inexpensive, and the ones that do win earn enough to support the whole race. The NSF estimates that the investment in fundamental research sees returns of 20 to 60 percent, even though thousands of horses never finish the race.

Supporting curiosity-driven science goes beyond a simple ledger statement. By publicly funding astronomy, an endeavor that attracts the "what good is it?" question like honey draws flies, Australia's CSIRO, the Netherlands Foundation for Radio Astronomy, and the United States' National Radio Astronomy Observatory VLA cultivated creative problem solvers. Public funding forged collaborations across disciplines and across borders to create a network that fostered cooperation, not competition. It wasn't just about exploding black holes. It was about outsmarting trillions of miles of intervening gas that would mask what they were looking for. It was about taking a clunky first shot and then refining and refining some more, something they were able to do only because they had the combined intellectual tools to attack the problem from several angles. These are precisely the sorts of people that world governments want educators to turn out en masse. Even without the multimillion-dollar royalties, astronomy's preparation of future problem solvers—along with their international community—would seem to be well worth the comparatively small investment.

✳ ✳ ✳ Now imagine a world where the theorizing about objects that might have existed at the beginning of time and then launching a search for these things were pursuits that were too fringe, too impractical. In this world, the development of WiFi lagged by just a few years, just long enough not to be of any use on the day of your heart attack. With your symptoms, you would have been rushed to the ER. And while it's true that you would probably have been surrounded by family and friends just a few days later, you likely wouldn't have been telling them how lucky you felt. Instead they would have been wishing that something could have saved you, never knowing that the decades-long international cooperation between theoretical astrophysicists, radio astronomers, and engineers was that something.

PART IV

✳✳✳ Pick Your Poison

IMAGINE. You're 22. You've got your whole life ahead of you. After you finish college, you're off to veterinary school. You'd also like to be a dog behaviorist, service dog trainer, and perhaps even take some of your furry friends to dog shows. The future looks bright. Then one day, out of the blue, you wake with severe pain and swelling on your left side. Doctors assure you that it's probably from an injury, and it will go away soon. The pain spreads ever so slowly. Within weeks, your entire arm feels like a giant raw nerve, the slightest touch excruciating. Your torso is next, and then your left leg, until you are no longer able to function without constant, searing pain. For a year, you go from state to state to find a doctor who can help, only to be met by the same puzzled looks and the same suspicions that you are making it up to acquire pain killers. Your hopes of becoming a vet must be put on hold. Probably permanently. Finally a diagnosis, a disease that has a name but no cure: complex regional pain syndrome. This particular disease has been de-scribed as the most painful experience on the planet, and you endure a pain rivaling childbirth 24/7. You learn quickly why this condition is also called the suicide disease. Medication doesn't help. You're allergic to morphine, and other medications either don't work or leave you in such a brain fog that you might as well not be alive. In desperation, you try a spinal cord stimulator—a surgically implanted device that connects electrodes directly to your spinal cord. Still no relief. Finally you hear of an experimental drug and you are approved by the doctor and insurance company to see if it can help. The list of adverse side effects scares you but not as much as the prospect of decades in near constant agony. The medication must be injected directly into your spine, but this is nothing compared with the spinal cord stimulator. Within hours, you feel better than you can ever remember feeling, and you even walk a few steps on your own before crashing back into your wheelchair. Twenty months later, you're actually taking your favorite dog to shows . . . and winning. Life is by no means "normal" for you, but at least you're living again. The thought of a young boy scouring the beaches of the Philippines, intrigued by a colorful snail that paralyzes its prey with a toxin-covered harpoon, never crosses your relieved mind.

24 DESPERATE TIMES, DESPERATE MEASURES

"No pain, no gain," goes the popular exercise motto. Watching Kerri Strug, Olympic gymnast, wipe out on her first vault in the 1996 finals, the world winced in sympathetic pain. But then she powered through the pain, limped to the runway, ran full speed ahead, tumbled head over feet in the air, and stuck what must have been the most excruciating landing in the history of the sport. Her coach, Bela Karolyi, scooped her up amid a chaotic scene that appeared to have been pulled straight from an inspirational movie. The world rallied around her heroism, as though she had single-handedly saved a thousand orphans from a burning school. For many, being told by their coach to save a thousand orphans would have been easier than doing what Strug was asked to do. In a world where the slightest twinge has us reaching for the ibuprofen, there is something superhuman in fighting through severe pain as though it isn't even there.

Perhaps displays like this, where pure willpower overcomes—at least temporarily—the instinct to curl up and nurse our wounds, have driven some cultures to involve extreme pain in their coming-of-age rituals. Young men of the Satere-Mawe tribe of the Amazon, for instance, prove their manhood by deliberately plunging their hands into pockets of woven leaves filled with bullet ants. The pain from the bites is agonizing, and the toxins cause convulsions, temporary paralysis, and extreme swelling. Some men even repeat this process time and again during their lives, despite having already proven themselves.

The concept behind deliberately enduring physical pain is fairly straightforward: physical pain provides us with a psychological weapon. If you can fight your way through dozens of ant bites (or a hundred razor blade cuts, or crucifixion with actual nails, or skewering muscle tissue and being suspended midair until the skewers tear through the flesh, or any number of extreme rituals that humans have devised to test their limits), you can fight your way through anything.

Vanquishing prescribed pain is a badge of honor.

Unscheduled pain is another matter, though, and cultures have always had a split personality when it comes to the pain life throws at us. In extreme rituals like those described above, it has evoked power beyond the merely physical. It has been spiritual, personal— possibly a punishment, possibly a purification. The ancient Greeks envisioned it as the workings of Poine, the spirit of retribution and vengeance, so it seems fairly clear where they fell on the "punishment/purification" question. Our English word "pain" derives from her name, in fact.

Other beliefs exhort us to embrace pain and suffering, inviting these companions cheerfully into our lives. Christianity itself was founded on the idea of the suffering servant, our daily pains uniting us with the divine. Pain is not meted out as vengeance but is instead a supernatural invitation to share in holiness, to cast aside the values of this world. It is interesting to note that the torture of criminals in the Middle Ages was seen not as some kind of sadistic revenge but as an act of mercy. The pain would help the criminal detach from his sinful proclivities, ensuring at his execution an express ticket to Paradise. Along a tangentially related vein are the ideas of Stoicism, which tells us that we cannot avoid pain, but we can refuse to allow it to rattle our inner peace; indeed, we can use it to strengthen that peace. Neither demonic nor divine, pain simply is a fact of life that we can face rationally.

Of course, not everyone has always agreed that pain should be an opportunity for personal growth. In hedonism and its modern counterparts, pain is to be avoided at all costs and eliminated whenever possible.

As purifying and well deserved as it might be, though, pain is
. . . well, a pain. So across the ages, the same cultures that have spir-
itualized pain have never tired of finding ways to combat it, but
where do you begin? In a world where so many things in nature can
serve to make us ill (or worse), it made sense to early humans that
perhaps some things in nature could serve to make us better. For
millennia, observations of other animals, mixed with a healthy dose
of trial and error, brought us to appreciate the healing properties of
various substances. Once writing took hold, it wasn't long before we
had transcribed a combination of knowledge, superstition, and wish-
ful thinking into cures and concoctions to rid our lives of suffering.

The first known prescription for what many agree to be a pain
remedy was apparently effective enough to merit being carefully recorded in a clay tablet over 4,000 years ago. The sufferer—or the
local Babylonian pharmacist—had only to do the following: "Purify
and pulverize the skin of a watersnake. Pour water over it and over
the amamashdubkaskal plant, the root of myrtle, pulverized alkali,
barley, powdered fir resin, the skin of the kushippu bird; boil; let the
mixture's water be poured off; wash the ailing organ with water; rub
the tree oil on it; let shaki be added." The exact identity of many of
these ingredients may have been lost to history, but the recipe gives
some idea of the lengths that people have gone to in their quest to
end pain.

One ingredient that has always been at the forefront of pain man-
agement is the poppy plant, from which an array of narcotic drugs
is derived. For all of recorded history, the poppy plant has been celebrated for its medicinal (and, let's face it, its nonmedicinal) qualities.
Known to the ancient Sumerians as "the joy plant," it has been cultivated, smoked, pulverized, crushed, crystallized, and concentrated,
for both its pain-killing abilities and its talent for inducing euphoria.
In 2012, scientists re-created an opium-based pain-killing paste prescribed nearly 2,000 years ago by Roman physician Galen. Fuscum
Olympionico inscriptum, "Olympic Victor's Dark Ointment," was
reserved specifically for those who had won an Olympic event. In
what amounted to an opium-laced transdermal patch, OVDO would

have provided both anti-inflammatory and pain-muting effects to the wearer. Given the religious underpinnings of the ancient games, during which sacrifices to Zeus would be made, it's notable that such an effective painkiller was provided only to the winners. Those Olympians who proved victorious in Zeus's own arena had earned their freedom from Poine, at least temporarily.

Another enduring warrior in our battle against pain is the willow tree. One of the earliest known records of its use is the 3,500-year-old Ebers Papyrus. Named for German Egyptologist Georg Ebers, the papyrus was a veritable physician's handbook, containing hundreds of remedies, most of which contained what is probably the all-time favorite painkiller: beer. The Ebers, along with another papyrus, was discovered in the Egyptology craze of the 1850s. For the paltry sum of 12 British pounds, Egypt enthusiast and self-taught hieroglyph translator Edwin Smith purchased the two scrolls and quickly found that they possessed a stunning wealth of information. One was named after Smith and became known as the Edwin Smith Surgical Papyrus, outlining in great detail various procedures that an ancient Egyptian surgeon should master. The other, which became the Ebers Papyrus after changing hands, was over 100 pages long, double-sided, and filled to the brim with medical knowledge. Transcribed around the year 1500 BC, the Ebers Papyrus seems to have been the ancient equivalent of a medical Wikipedia, a single source containing the combined healing methods from the previous millennium or more.

What the Ebers Papyrus tells modern scholars is that, like many cultures across the globe, the ancient Egyptians had some idea of anatomy and physiology mixed with a large portion of supernatural gap fillers. Their medicines were surprisingly varied, taking advantage of the plants and animals at their disposal, but their understanding of the fundamental reasons behind ailments was unsurprisingly sketchy. They felt that most illness could be traced to an excess of a substance called *wekhudu*, one of four essential components of the body alongside water, blood, and air. Many of their "cures" involved driving this excess from the body. This same basic philosophy of balanced substances found its way into Greek thinking more than 1,000 years later, when Empedocles suggested that there are four essential humors in the body—blood, yellow bile, black bile, and phlegm—each corresponding to one of the four universal elements air, fire, earth, and water. In both cultures, removing the excess, often by bloodletting or purging, was the prescribed treatment for many ailments. More extreme practices included trepanation, literally drilling a hole into the patient's head, a treatment that was supposed to somehow balance the humors, relieve pressure, release the bad spirits, or something. It is amazing that a large percentage of those who underwent trepanation managed to survive, living out the rest of their lives with a quarter-sized hole in their skulls. It is even more amazing that there is a modern, albeit extremely small, movement that advocates the return of trepanation as a means to keep the brain young and vibrant.

Although many treatments described in the Ebers Papyrus have been discarded as our understanding of physiology has progressed, there is one notable substance that it and other ancient texts laud. According to the papyrus, should someone come to the physician with a hot, swollen infected area, the physician "must make cooling substances for him to draw the heat out [using] leaves of the willow." Elsewhere in the 877-paragraph document are further references to the pain-relieving abilities of the bark and leaves of the willow, or tjeret, as they called it.

Now medicine from the willow bark goes by the name aspirin.

25   BARKING UP THE RIGHT TREE

The Ebers Papyrus was by no means unique in its recommendations for willow extracts. More than 2,400 years ago, Greek physician Hippocrates, for whom the Hippocratic oath is named, recommended a tea brewed from its bark to help with the pains of childbirth. For centuries, the Chinese prescribed a willow derivative for pain and fevers, and many Native American and African cultures were no strangers to the analgesic and antipyretic powers of this common tree. It was known, forgotten, and rediscovered throughout the centuries, but in 1763, it was finally discovered for the last time by Reverend Edward Stone, an Englishman. In a daring letter to the Royal Society, this nonscientist described "An Account of the Success of the Bark of the Willow in the Cure of Agues," explaining,

> There is a bark of an English tree, which I have found by experience to be a powerful astringent, very efficacious in curing aguish and intermitting disorders.
>
> About six years ago I accidentally tasted it, and was surprised at its extraordinary bitterness; which immediately raised in me a suspicion of its having the properties of the Peruvian bark. As this tree delights in a moist or wet soil, where agues chiefly abound, the general maxim, that many natural maladies carry their cures along with them, or that their remedies lie not far from their causes, was so very apposite to this particular case.

He is mistakenly referred to as *Edmund* Stone at the beginning of the letter, indicating that the Royal Society was confusing him for an actual academic by that name, a mistake that likely prompted the scholarly society to take Reverend Stone's treatise seriously.

The aguish disorders he refers to are maladies marked with fevers, body aches, and chills. Agues were often synonymous with malaria, which literally means "bad air." The thinking of the time was that the stagnant water produced a toxic vapor—a bad air—that caused listlessness, fever, and often death. The notion that a lowly mosquito, thriving in the still waters, might carry an invisible being that infected its human host was over a century away. All anyone at the time knew was that the agues coincided with the dampness. The bitterness in the bark, which Stone nibbled for reasons unknown, was reminiscent of a Central American tree (cinchona, or quinine) that was known to be a successful treatment for malaria, and the link between the two was immediately forged in his mind. Fortunately, it's likely that most of Stone's guinea pig patients did not genuinely suffer from malaria, and so the lack of quinine in willow bark did not come back to bite him. Instead, what he found was that dried, powdered willow bark brought a great measure of relief when administered every four hours or so to someone suffering either a fever or aches and pains. But why?

For Stone, it was the doctrine of signatures at work. Since ancient times, there had been a persistent belief that the "signature" or shape of a plant held the key to its curative ability. Thus a plant whose flowers looked like blue eyes held the cure for eye ailments. The toothwort flower could help with dental problems. The idea of the doctrine of signatures had been resurrected with the work of Paracelsus, a man whose name sounds distinctly Roman but who was in fact a sixteenth-century German whose given name was Philippus Aureolus Theophrastus Bombastus von Hohenheim. The original Celsus, a first-century BC physician, had lived in Rome, and von Hohenheim's adoption of Paracelsus was meant to suggest that he was at least on par with this ancient authority.

While the original doctrine of signatures looked at only the physical shape of the "cure," it evolved over the centuries to encompass a more environmental view. If the tooth-shaped flower held the cure for the tooth, then the bogs associated with malaria would hold the cure for malaria. Or so Stone reasoned. But his willow cure was a bitter pill to swallow, quite literally, and the subsequent purification of willow's active ingredient didn't help matters any. By the 1820s, chemists (no longer alchemists) had managed to isolate a handful of substances that impact our health, for better or for worse: caffeine, strychnine, and quinine, to name just a few. Several groups of chemists had a go at the willow bark, successively improving the purification of the bitter ingredient. The extract from the willow— or *salix*—tree became known as salicylic acid. The same substance was rediscovered independently by German chemist Karl Lowig, who called his finding *spirsaure*, after the meadowsweet plant from which he extracted it. As it turned out, salicylic acid was found in a host of plants, a convenient feature for researchers. But it was acidic and once purified was reportedly so unpleasant to ingest that most people would sooner suffer fevers, aches, and pains rather than take a second dose. If only, chemists thought, there were a way to keep the analgesic and antipyretic properties and lose the unpleasant bitterness and caustic effects.

Advances in chemistry, along with a healthy dose of curiosity and old-fashioned luck, led to the unlikely introduction of coal tar residues in the race to manage pain. English chemist William Henry Perkin had lived his life experimenting on pretty much any substance available. With the industrial revolution in full swing in the 1850s, he had access to the byproducts of coal manufacturing, which, as far as anyone could tell, were fairly useless. His curious fiddling led to an impressive new purple dye—mauve—that catapulted Perkin into the wealthy class almost instantly. The coal tar experimentation led to other new, vibrant dyes, and soon textile manufacturers around Europe were keen to cash in on its unexpectedly colorful yields.

One such company was cofounded by Germany's Friedrich Bayer, who helped establish a growing business that produced new dyes using cutting-edge chemistry, meanwhile dumping untold amounts of toxic waste into the local streams and rivers. The brand name Bayer is probably familiar to you, but not for its dyes. The Bayer company, recognizing the commercial potential of young scientifically minded graduates, had a division that these days would be termed R&D. In those days, and at that company, it was a proving ground for the glut of chemistry graduates who were hoping to be part of the coal-tar derivative bubble. For a year, newly minted chemists could endure a low-paying fellowship with the company, and, if they showed promise, they might be offered a permanent, higher paying job. Carl Duisberg was one such chemist, and his skill at creating new dyes was noticed. While he was a talented chemist, what really set him apart was his business acumen.

Bayer's foray into pharmaceuticals began when the coal-tar derivative acetanilide was accidentally administered to a local man suffering from intestinal parasites. Happily, the pharmacist's mistake did not kill the patient. The cause of the grievous mix-up has never fully been settled, but, as luck would have it, the substance the patient ingested possessed great fever-reducing powers. A rival manufacturing company promptly seized the opportunity, purified and patented the active ingredient, and profited richly off a mistake that would have launched a lawsuit had it occurred in our century. The substance became known as Antifebrin, with an active ingredient that would be further modified until it became what we know as acetaminophen, often referred to by the brand name of Tylenol.

Duisberg saw no reason that the Bayer company, whose corporate ladder he'd successfully climbed, couldn't do the same, and so he launched a pharmaceutical division. Waiting for a pharmacist's fortuitous blunder, though, was too chancy, so he put his money on a surer thing. Since nobody had yet patented salicylic acid or figured out how to reduce its unpleasant side effects, Duisberg made formulating and marketing the pain killer the number one priority of Farbenfabriken vormals Friedrich Bayer and Company (the Dye Factory Formerly Known as Friedrich Bayer and Company). As recounted in Diarmuid Jeffreys's book *Aspirin: The Remarkable Story of a Wonder Drug*, the pharmaceutical side of the Bayer company appeared to be little more than an attempt to re-create a chef's secret recipe: "Different combinations were tried out; something was added here, something reformulated there. Sometimes they came up with a promising lead; more often than not, their trials ended in failure . . . then, perhaps by luck or ingenuity, [they would] stumble upon a useful product."

Such a description sounds remarkably similar to the process of science, so much so that one might wonder why we don't simply leave this job to private industry and take universities and governments—and our own money—out of the loop entirely. Although all of the testing, tweaking, and refining has the ring of science, Bayer had a specific goal in mind. Their work was not driven by curiosity but by potential profits from a specific product. However, that specific product *did* have its origin in a curious taste of willow bark. Moreover, had Bayer failed in its quest, it is unlikely that you would know their name at all. Yet scientific failures—like the experiment of Michelson and Morley on the relative motion of matter through the ether or the O'Sullivan, Ekers, and Shaver search for exploding black holes—are often just as valuable as scientific success stories. In fact, there has been a recent push for journals to publish more papers with null results, partly to keep other researchers from wasting time reproducing those experiments, but partly because finding nothing tells you something. In science, the process of understanding Nature is what matters, and every result provides new insights that might hold the key to a breakthrough down the road.

Obviously, Bayer succeeded in its quest to develop an effective painkiller. In 1897, Bayer's most famous useful product—acetylsalicylic acid, otherwise known as aspirin—was created. With all the painkilling properties of salicylic acid but without the corrosive acidity that burned the esophagus and irritated the stomach, Bayer's new creation was directly marketed to tens of thousands of physicians so that doctors could prescribe it by name. Ironically, pharmacists, or chemists, of the time would strive to dispense only what the physician had ordered; even in the 1890s, generic equivalents were suspect, although apparently it wasn't unheard of to provide the patient a completely unrelated substance leftover from an industrial process. Now, over a century later, Bayer's marketing success is clear: almost nobody is aware that the company started in the dye-making business.

Despite a tongue-twister of a name, acetylsalicylic acid is actually a fairly simple molecule, all things considered. Salicylic acid in and of itself consists of a six-carbon benzene ring, with two additional molecules jutting out of the ring, giving the chemical structure the look of a hexagonal alien head with two short antennae. One of the attached pieces contains an oxygen atom and a hydrogen atom—OH—and it is this piece that produces the ugly side effects. What the Bayer chemists did was to change this antenna to a different molecule, one related to another common substance. Their molecular substitution is why your old aspirin might occasionally smell a bit like vinegar. The airborne water in a humid bathroom can break aspirin down into acetic acid (vinegar) and salicylic acid (the pain-killing ingredient). Your degraded aspirin won't kill you, but it will certainly help you understand why patients in the 1800s were reluctant to take the willow extract unless they were in extreme agony.

26   THE INSIDE STORY ON PAIN

It's one thing to isolate the active ingredient in aspirin and other painkillers and another thing entirely to understand why they alleviate pain. Adding to the mystery is the problem of why the body senses pain to begin with. The happiness of a patient, though, lies in the doctor's ability to stop the pain, not explain it. People will happily chew on willow bark if it quells the pain of arthritis, and the chemical structure of the active ingredient or the mechanism of its action never crosses their minds.

For many types of pain, a class of biochemicals known as prostaglandins seems to be the culprit. These 20-carbon molecules are chemical messengers that, among other things, tell the neurons in your spine to transmit pain and adjust your body's thermostat, causing you to have a fever. It might seem as though getting these things out of our body chemistry would be a good idea, except for the fact that they also help regulate kidney function, protect the lining of the stomach from its own acid, and activate blood platelets in clotting. Clearly these are vital functions that we don't want to sacrifice to get rid of a headache or a fever. But many of us choose to temporarily override these functions in order to get relief, so we ingest aspirin or other nonsteroidal anti-inflammatory drugs (NSAIDs) to inhibit production of prostaglandins. The downside to long-term habitual use of such NSAIDs is that the body no longer has the stomach lining protection it once had, and ulcers or even internal bleeding can result. Left unchecked, this hemorrhaging can result in death, which is why warnings on ibuprofen bottles tell users to be on the lookout for symptoms.

Pointing the finger at prostaglandins, though, is simply passing the buck. Why would blocking them (or, indeed, blocking or transmitting any chemical) ease pain at all? To understand the sensation of pain requires digging deeper into the things that send sensory messages through our bodies. Roman physician Galen, who gave us the prescription for the clever opium-laced pain-killing patch, spent a fair amount of time trying to understand our inner workings. What he found was a network of fibers that spanned the body and seemed to converge in the spinal cord and ultimately the brain. As far as Galen could tell, these fibers were responsible for both sensation and motion, appearing to be hollow tubes through which the "animal spirits" were transmitted.

The idea of animal spirits was as tenacious as the doctrine of signatures and held weight in scientific circles well into the eighteenth century. By then it was the more fashionable "animal electricity" that was coursing through a system of nerves, a word that meant merely sinews or fibers, a nod to their visible appearance. Italian scientist Luigi Galvani possessed the perfect combination of interests to explore the nervous system. As a philosopher, physicist, and physician, he understood both the external and internal physical worlds and had the ability to spot commonalities.

While watching his wife prepare dinner one evening, Galvani noticed that the muscles of frogs' legs twitched upon contact with the knife, even when severed from their owners. Naturally, he followed up on this observation, hanging frogs' legs from metal hooks and touching them with metal probes, whereupon they obediently twitched. Animal electricity seemed to be the only possible conclusion, but his friend, physicist Allesandro Volta, begged to differ. The twitches were the result of electricity, yes, but not from some special animal variety. So, while Galvani worked to prove that he had managed to demonstrate animal electricity, Volta instead used the knowledge that something in animal tissue can cause electrical current to flow between two different metals. Galvani clung to the idea of an electrical animating force, while Volta instead applied the new knowledge to create the first battery.

As it turns out, electricity is the key to pain, pleasure, movement, and even thought, but this understanding did not come easily or quickly to scientists. Now, more than two centuries after Galvani's celebrated jumping frog legs, we have a picture of the nervous system that at the most fundamental level looks like a schematic for a highly complicated circuit. The concept of animal electricity has given way to the idea of a sterile flow of charged particles across bridges and through gates.

On the cellular level, it is our neurons that convey electrical signals throughout the brain and body. The brain alone contains about a hundred billion neurons, each connected to thousands of other neurons in the brain. Neurons come in a variety of shapes and sizes—some as long as a meter, and some less than a millimeter in length. The standard neuron sketch looks a bit like a great willow tree, complete with an elongated trunk, roots, and branches. The branches, or dendrites (a word that is merely Greek for "treelike"), receive signals from other neurons. These signals are then transmitted through the trunk-like axon and sent via neurotransmitters out the axon terminal and across the synapse to the next neuron in line.

Although there are thousands of types of neurons, the mechanism for their behavior is basically the same. The membrane of the neuron's dendrites has gateways through which charged particles can flow. The membrane itself is electrically polarized, the outside with a slightly more negative charge than the inside. What this means is that positive charges are clamoring to get in to balance everything, but the gatekeepers keep the doors closed. The charges at work are ions of potassium and sodium, and there is a special sodium-only gate through which the sodium ions can flow and another chemically guarded gate that allows only potassium ions to flow.

In the event that the chemical gatekeepers have been ordered to open the gates, sodium ions will rush into the neuron's dendrites like shoppers on the opening day of a sale. The interior quickly becomes far more positively charged, and this charge imbalance triggers another chemical gatekeeper to allow potassium ions out of the neuron, balancing the situation. The influx of positive sodium ions followed by the almost instantaneous outflow of potassium ions causes an electrical blip that travels down the neuron, almost like the flick of the end of a rope causes a wave to propagate all the way to the other end. At speeds of up to about 500 kilometers (300 miles) per hour, the signal shoots through the axon. When it reaches the other end—the axon terminal—the signal is shipped across a synapse to the next neuron. That signal will then tell the next round of gate-keepers to open or close their gates, and thus the signal propagates from one neuron to another.

Some types of neurons—particularly the pain-sensing ones—have exceedingly stubborn chemical gatekeepers. Springing into action only when an incredibly high threshold has been met, these so-called nociceptors are the neural equivalent of a hot line to the brain, informing it that something has been burned or cut or subjected to some other kind of damage. In 1906 physiologist Sir Charles Scott Sherrington proposed that nerves sensitive only to noxious stimuli existed. His Nobel Prize–winning work on neurobiology also earned Sherrington an invitation to a distinguished Yale College lecture-ship. He was only the second person to receive this invitation, the first being J. J. Thomson, the discoverer of the electron. Sherrington's basic hypothesis was that pain exists not as a punishment and not as purification but as a message from the body to the brain that something has been damaged. Nociceptors sensitive to heat, for ex-ample, could tell you quickly that your hand was too close to the fire, and the unpleasantness of the pain would compel you to move your hand. Those same neurons, though, would be unperturbed by, say, a light caress that would trip the gatekeepers of other sensory neurons.

While a logical proposition, the existence of nociceptors would have to wait until the 1960s to be verified. Since then, scientists have found out that nociceptors are not a one-stimulus-fits-all type of neuron. Some are attuned to thermal damage (and even the capsa-icin in hot peppers), while others can tell by the presence of certain molecules in the vicinity that a nearby cell has been damaged and leaked those molecules into the area. Like other sensory neurons, nociceptors send their information to the spinal cord, which is to the nervous system what the Memphis superhub is to Federal Express. Pretty much everything is received, rerouted, and shipped out from there.

So what does the bark of the willow tree do to keep these nociceptors from sending their messages to the brain? As it so happens, one of the chemicals released when cells are damaged is quickly converted into prostaglandins. Those prostaglandins keep the potassium gatekeepers from allowing potassium ions out of the neuron after the sodium ions have invaded, a blockage that serves to make that nocicepting neuron touchier and the pain more easily registered. By blocking the prostaglandins, aspirin raises the breaking point for the nociceptor gatekeepers, which means the headache stops aching quite so much.

Unfortunately, aspirin won't do a thing for you if you're stung by a cone snail. And yet, a cone snail might ease your pain in the most unexpected ways.

27 A BITTER STING

Anyone who has ever been stung by a cone snail and lived to tell the tale will agree that nothing could be farther from a painkiller than this benign-looking beast. They come in a number of varieties—*Conus geographus* (the geography cone), *Conus magus* (the magician's cone), and *Conus cedonulli* (the matchless cone), to name just a few. The average cone snail is the size and general shape of a rosebud, but its thorns are far worse. Their colorful shells adorned with complicated patterns make them a favorite among shell collectors, assuming the collectors are fortunate enough to find uninhabited samples. So prized were these shells in the eighteenth century that a Vermeer painting was auctioned for a fraction of the price of a single *Conus cedonulli* shell.

But the cone snails are far more than just a pretty shell. On the off chance that a casual diver runs across one of these creatures while enjoying a tropical getaway, he will soon find out that this minute mollusk can take perfectly good care of itself. The moment it senses that it is in danger, the cone snail is likely to launch a tiny harpoon, certainly not a typical feature of garden-variety snails or hermit crabs. The barbed harpoon will sink into the meat of the hand. In some cases, pain more severe than a hornet's sting registers immediately as the nociceptors tell the brain that there's been tissue damage. In others, there is a puzzling lack of sensation.

Pumped into the harpoon is the cone snail's venom, a complicated toxic cocktail for which humans have devised no antidote. Depending on which of the 700 species of snail has been handled, tingling and numbness might follow the pain or prevent it entirely, first in the area around the sting and then sweeping up the entire limb. This might initially seem to be a plus, given that the pain yields to numbness, but it only reinforces the fact that this is no ordinary snail. Fainting, paralysis, coma, or even death can result. One of the larger species of cone snail—the *Conus geographus*—is nicknamed "the cigarette snail," not because of its size or shape but because it is said that the victim has about enough time to smoke one last cigarette before dying. (This story is most certainly apocryphal, say cone snail experts; it takes hours for the victims' diaphragms to become paralyzed, thus causing death. This is cold comfort, however.)

Found in every tropical ocean on Earth, cone snails are fearsome hunters, and fully 70 species are successful fishersnails. It simply doesn't appear possible. A sea snail catching a quick, darting fish would seem as likely as a garden snail snagging a hummingbird mid-flight. But through an intricate system of luring and harpooning, the cone snail can catch lunch in the blink of an eye. Some fishersnails dangle a wormlike proboscis outside their shells, enticing a hungry fish to investigate. When the fish gets too close, the snail launches a harpoon into it.

A cone snail's harpoon is an amazing biological development. The head of the harpoon contains a disposable barbed "tooth," which the snail restocks after every harpooning. The barbs keep the harpoon in the fish while the snail squeezes its potent venom from its venom sac through the line of the harpoon and into its victim. The fish will thrash momentarily—less than two seconds, usually—and then assume a rigid catatonic state with all its fins jutting outward. The otherworldly mouth of the snail then envelops the entire fish with a soft funnel of tissue, until all that can be seen is the still-paralyzed fish tail. Within hours, the bones and scales are spat back out, the cone snail sated and the fish reduced to a small pile of debris on the seafloor. As far as scientists can tell, cone snails have been dining in this fashion for about 20 million years.

Although a young child might be horrified to see a clownfish dispatched in this way ("It's eating Nemo!"), cone snails are largely not an issue for humans (unless you consider their use as an essential ingredient in a Filipino soup). Although stings from the geography cone have a 70 percent fatality rate if left untreated, only a few dozen deaths attributable to cone snails have been recorded over the past 150 years, far fewer than the number of deaths attributable to cows (about 20 per year). And yet the fact that this tiny creature can kill a human at all is perplexing, and, fortunately for many people suffering chronic pain, this lethal ability perplexed the right person.

28   A SIMPLE QUESTION

In 1941, fear gripped the world. Fred Griffith—who we learned in part II discovered, by injecting various strains of pneumonia bacteria into countless mice, that there was an interchangeability to inherited traits—died in one of many bombing raids on England that year. As the year drew to a close, the attack on Pearl Harbor pulled the United States into an ever-growing global war with no end in sight. The Philippines was not immune to the worldwide suffering. The day after Pearl Harbor, the Japanese forces devastated a United States Air Force base north of Manila.

It was a dangerous time to be born, the Philippines a dangerous place, but Baldomero Olivera had no say in the matter. Nicknamed "Toto" because a young cousin couldn't pronounce the more common pet name "Totoy," he turned five the same year that this island nation gained its independence from the United States. The tropical climate of the Philippines perpetually beckoned the children to be outside where nature gladly provided the entertainment. Toto, like nearly everyone else who has walked along the beaches near his hometown of Manila, was drawn to the patches of color that broke up the pale sands. Seashells. Oranges, pinks, golds, browns, purples, whites, fan-shaped, swirled, spotted, striped, elongated, shiny, spiked, corrugated . . . the shells on the beach were as diverse as the marine life near this archipelago. When he happened upon an unknown specimen, he would take it home and try to find its match in his well-used library of marine life. It was a pastime that he never outgrew.

Toto never outgrew his nickname, either, or his childlike inquisitiveness. His curiosity led him to excel in school, graduating summa cum laude in chemistry from the University of the Philippines, and ultimately across the Pacific where he studied biochemistry at the California Institute of Technology. Borrowing a page from Arthur Kornberg's book of interesting biological molecules (see part II), Toto discovered one of the enzymes that could join together two DNA molecules. Indeed, this seemed to be the brightest, most obvious path for a promising biochemist. But after a postdoctoral stint, Toto wanted to return to his home country of the Philippines, where he was welcomed with open arms and an empty lab. There was simply no way he was going to pull together a sophisticated enough lab to study DNA replication and enzymatic breakdown, certainly not a competitive one. For the industrious and ever-curious scientist, though, there is never a shortage of potential experiments.

"I wanted a project that had some local advantages and required no equipment at all," Baldomero recounted, holding the mauve-and-white patchwork shell of a *Conus geographus*. It is about the size of a small pear, its beauty deceptive. His revised research goal was to answer one simple question: "Why is this snail able to kill people?"

In a room that appeared to be little more than a garage, with an experimental setup that looked like an eighth-grade shop project—a saddle of wire mesh suspended over a plastic rectangular trough— Toto and his students began studying the deadly snails. To hear him describe the initial experiment, a cynical listener would be certain that nothing useful was ever going to come from this. The basic procedure was as follows: hang a mouse on the underside of the mesh, inject it with a dose of cone snail venom, and then see how long it takes the mouse to become paralyzed and fall into the bucket. The number of mice injected began to rival even Fred Griffith's years-long *Pneumococcus* experiments, but this odd mouse freefall device helped them isolate the two main components of the snail's venom that caused the paralysis.

The first component separated out was dubbed α-conotoxin (the Greek letter alpha) G1 (gee one). Biochemically speaking, it was a fairly small molecule. Consisting of 13 amino acids strung together in a type of molecule called a peptide, the α-conotoxin appeared to have the same effect as cobra venom, blocking the communication between the nerves and the muscles at the neuromuscular synapses. Neurons speak to muscle membranes in much the same way that they talk to each other. At a neuromuscular junction, calcium ions are shuffled around when the voltage at the nerve ending is altered. The calcium gatekeepers allow the positively charged calcium ions to flow into the nerve ending, and this change triggers the release of special chemicals called neurotransmitters that cause a mechanical change in the muscle. By blocking the signal from nerves to muscles, this toxin created a firewall between the brain and the muscle so that instructions for the muscle to contract were simply not received.

The other paralyzing component was known as μ-conotoxin (with the Greek letter mu) and was similarly small—22 amino acids in length. Toto and his students found that the second chemical had a different mode of attack. Instead of blocking the signal from the nerves to the muscles, it simply wipes out the signal down the muscle membranes themselves. This toxin allows the signal to get from the brain to the muscle but leaves the muscle completely unable to respond. Fugu, or puffer fish, have a similarly acting toxin. A small nibble of the wrong fugu tissue will kill you, so it is only from the hands of the most skilled (and correspondingly highly paid) chefs that you want to receive this Japanese delicacy. With two paralyzing toxins in its arsenal, a cone snail sting is described by Olivera as equivalent to "being bitten by a cobra and eating a lethal dose of fugu at the same time."

You might think that finding the answer to the research question posed would be a scientist's dream, but the truth was that Olivera found the results almost anticlimactic. The mechanisms of the two types of toxins they uncovered had already been established using the toxins from snakes and puffer fish, and it seemed like there was little else to do with the cone snails or the sparsely equipped lab. So Olivera returned to the United States in 1972, with the hopes of getting back into research on DNA enzymes in a competitive lab at the University of Utah.

Had he followed his own scientific instincts, his work might not have found its way into a discussion of pain relief. But it's the rare scientist who works in isolation, and the innocent curiosity of a student turned his attention back to the cone snails. What would happen, his undergraduate Craig Clark asked, if the toxins were injected directly into the central nervous system instead of into the body cavity? And what do the other components of the cone snail venom do? As far as Olivera could tell, there were two active ingredients among a buffet of otherwise inactive peptides, but Clark persisted. He wanted to explore those other components.

"I discouraged him from doing this," Olivera said. "But the reason I think why most creative research is done at universities rather than anywhere else is because the students do what they want. They don't necessarily listen to their professors, and that's a very good thing."

Clark forged ahead, separating out literally dozens of peptides from the venom and injecting them intracranially into endless batches of mice. The result was as surprising as being on the receiving end of a snail's harpoon.

Mice were no longer simply being paralyzed. Some were scratching, trembling, jumping and turning. Some were walking in circles. Some were convulsing. A particular compound would put the youngest mice to sleep for hours while causing older mice to pace and climb frenetically. The cone snail venom turned out to be a virtual apothecary's shop.

The venom also was bringing to mind the words of Paracelsus and his modified doctrine of signatures: "Alle Ding' sind Gift, und nichts ohn' Gift; allein die Dosis macht, daß ein Ding kein Gift ist." ("All things are poison, and nothing is without poison; only the dose permits something not to be poisonous," or more simply, "The dose makes the poison.")

It's interesting to note that the German word "Gift" translates into "poison." In the case of the cone snail, it also translates into "gift."

29   THE GIFT OF THE MAGUS

"As so often happens in science," Olivera would later explain, "while we were addressing the problem we initially wanted to address, we discovered something much more interesting." Having fully answered the question they originally posed, Olivera and his students now had scores of new questions. With hundreds of species of cone snail, and over 100 unique peptides produced per species, the possibilities for research seemed endless.

Predictably, though, extensive research into the compounds in cone snail venom requires, well, venom, which is not always easy to acquire in a controlled manner. The cone snail has only a small pouch of venom that it pumps through a duct into the disposable tooth of its harpoon. For some cone snails, an adequate sample of venom could be taken during a dissection, which obviously necessitates a dead cone snail. Other cone snails, though, had venom sacs that were quite small. Olivera was hoping to avoid needlessly killing so many of them simply to get a workable sample, and he got the strange notion to "milk" the snails instead. With captive snails in an aquarium, he hoped to extract enough venom to study while leaving his specimens alive and available for further venom donations. Unfortunately, milking a cone snail was a tricky proposition, and no students were stepping up to volunteer. Then came Chris Hopkins.

Chris Hopkins, an undergraduate student, seemed completely unfazed when we suggested the snail milking project to him. He said he wanted to think about it for a few days, came back and asked for a little cash. He returned with a box of condoms and started blowing one up. He rubbed the inflated condom against a goldfish and then lowered it into a tank with hungry *Conus striatus* buried under the aquarium sand. Several came up from the sand, and one harpooned the condom with such force that Chris let go in surprise. The condom floated up with the snail still attached to it through its harpoon and proboscis. The sight of an inflated condom floating at the surface, with a tethered snail swinging like a pendulum below it, was one of those moments that should have been recorded with a camera.

Perhaps it's best that the visual is left to our imaginations. It seems when scientists record surprisingly amusing events on camera, the public is all too willing to release its own venom. Surely using public money to purchase condoms to bait harpoon-shooting snails would have been in the running for a Golden Fleece award had Proxmire gotten wind of it. What it told Hopkins, though, was that the snails could be fooled into injecting their venom into a collection vessel, and after hundreds of harpoonings (and an improved collection vessel), Olivera's team managed to amass enough venom to study in detail.

One of the more curious compounds from the *Conus magus*, or magician's cone, was cleverly dubbed the "shaker peptide" because of its ability to make mice shake uncontrollably. Given the similarity of this behavior to epileptic seizures, it seemed like a potentially valuable peptide to explore. And so in the early 1980s, Toto's undergraduate student Michael McIntosh decided, from the tens of thousands of conus peptides, to concentrate on this one.

McIntosh had not originally set out to spend his summer studying conotoxins or mice or anything biological, actually. Fresh out of high school, he had really wanted to be part of a touring theater group (he was supposed to play Motel the tailor in *Fiddler on the Roof*), but his friend Craig Clark talked him into joining the cone snail lab. By this point, Toto had completely switched gears from his previous (successful) pursuit of DNA-digesting enzymes, largely due to the urgings of Clark, who apparently could persuade a scientist to change his research focus and a theater aficionado to become a biologist.

McIntosh succeeded in isolating and purifying the shaker peptide, which was given the name ω-conotoxin (the Greek letter omega) MVIIA. Follow-up work by neuroscientists at the University of Utah showed that the molecule blocked one of the calcium ion channels between the neurons and the muscles. The calcium ion gatekeepers are the ones allowing the neurons to talk to the muscle, but, even among this subclass of molecular gatekeepers, there are subtle differences. Some calcium channel blockers are not picky; they'll block all the channels. This conotoxin, though, acted only on one, and it happened to be the very one that let the brain know about pain. What this means is that your arm could be plunged in a vat of boiling oil, nociceptors firing away frantically, but that message simply would not get past the firewall in the spinal cord.

It didn't take long for the implications of this finding to sink in. If this particular conotoxin could be injected directly into the spinal cord, then pain would fail to register.

And thus was born Prialt, a *pri*mary *alt*ernative to morphine, so named because, as Olivera points out, "They couldn't continue to call it ω-conotoxin. You don't want to take a *toxin!*" Also known as ziconotide, Prialt was developed as SNX-111 in 1993. It underwent over a decade of testing and was finally approved by both the European Union and the United States in 2004. Prialt is used almost exclusively to treat excruciating long-term pain associated with nervous system disorders, AIDS, and cancer. Unlike pills or simple injections, this drug is sent directly into the cerebrospinal fluid in the space that surrounds the spinal cord with the surgical insertion of a special pump.

It might seem like a miracle discovery, but there is a veritable menu of unpleasant side effects that drive people to seek an alternative to morphine's primary alternative. For some, though, the hallucinations, paranoia, the desire for self-harm, dizziness, fainting, and a loss of memory and balance that can result from taking Prialt are worth the risk because of the drug's ability to end severe pain. Fifty times more potent than morphine, Prialt is nonaddictive, and the body does not build up a tolerance to it.

The way Prialt is able to sidestep the problems befalling morphine is that it skips the middleman, neurologically speaking. Morphine stimulates something called an opiate receptor. This receptor will in turn act on the calcium ion channels. Unfortunately, that receptor becomes essentially habituated to the stimulation, so after several days of morphine treatments, the pain will return. Prialt, in contrast, goes straight to the calcium ion channels, which are slaves to a much more basic physical process at the atomic and molecular level. They can't become resistant to it any more than a proton can become less positively charged. For sufferers of long-term, intractable pain, the poison that became Prialt is truly a gift.

"Snails are ultimate practitioners of combination drug therapy," quips Olivera. And with tens of thousands of peptides to study, cone snails might hold the answers to hundreds of pharmacological questions that were nowhere on his radar when he was attaching mice to the bottom of some wire mesh. "We weren't looking for a drug," he reflects. "We were just trying to understand what's going on with these snails." As for the value of impractical research, "I think the lesson is that if you're trying to understand basic science, very often you stumble on things people didn't know and that have applications that you never expected."

* * * Now imagine . . . no, you don't want to imagine a world where the pain never ceases. Instead you focus on this one. Night falls, and your best friend is thumping his tail near your feet, thanking you for another day of simple joys. Your pain has become so well controlled that sometimes people don't even notice you have a disability. You smile as you wonder whether you can take that tropical vacation you've always dreamed of. After all, you have the perfect shelf to display some shells.

PART V

✱✱✱ Dreaming of the Star Treatment

IMAGINE. You finally have some free time and you think you'll catch up on some reading. Your friends have recommended Rebecca Skloot's *The Immortal Life of Henrietta Lacks*, so you crack it open one lazy afternoon. It seems like a great choice, something that digs deep into the issues of social strata, medical ethics, and racism both before and after the civil rights movement. Good food for thought. But as you read, the personal story of Henrietta Lacks begins to hit you. In what seemed a medieval form of torture, the treatment for cervical cancer in 1951 was literally to sew radium inserts inside the 30-year-old Lacks, with the hopes that the radiation from them would kill the aggressive cancer before killing the patient. The side effects were incomprehensible. The region near the radium inserts became cooked with radiation, causing searing pain. The skin of her abdomen was blackened from the treatment. She was weakened and nauseated, all while trying to take care of five children, one born just months before the cancer diagnosis. You read with horror of her ignoble death and the unmarked

grave of a young cancer victim who, even in death, suffered indigni-
ties that would have legal and ethical ramifications in the twenty-first
century. Surely things have changed since then, you think, recalling a
colleague's months-long absence during her own battle with cancer.
Surely in this age of smartphones and smart bombs we have progressed
past the scorched earth strategy for cancer treatment. Haven't we? You
take a quick glance at the National Cancer Institute's website only to
find that, while many things have improved in the intervening decades,
some cancer treatments still seem barbaric and outdated. You feel a
wave of anger, sadness, and helplessness, and you wonder if perhaps
it's time for the nation to divert some of its monetary and intellectual
resources from the impractical and pointless studies about impossibly
distant objects to the problems we face right here. After all, it seems
ridiculously far-fetched to think that funding attempts to understand the
precise inner workings of our nearby star or the chaotic surroundings of
unimaginable objects could ever be of any practical use to someone you
know with cancer. It doesn't dawn on you that stars are our cousins, our
atoms forged together in the same furnaces, and that by understanding
them, we can understand ourselves.

30   INGREDIENTS OF THE STARS

In 1806, Jane Taylor, a somewhat obscure English poet, penned the most famous astronomical opus of all time. Perhaps you are familiar with it. If not, it's a safe bet that every child in your neighborhood is. Let me refresh your memory:

> Twinkle, twinkle little star
> How I wonder what you are.

Perhaps it was 30 years of hearing this song that drove French sociologist Auguste Comte to declare of stars, "We understand the possibility of determining their shapes, their distances, their sizes and motions, whereas never, by any means, will we be able to study their chemical composition." This was a particularly interesting statement for a nonscientist to make, especially considering that in 1835 the universe had already begun dropping subtle hints that a star's light might reveal telling information about the nature of the star.

Extracting information from starlight required dissecting it first, separating the colors by means of a prism. Only then could one even begin to divine the fundamental workings of the stars. In 1802, just a few years before the first publication of Taylor's immortal poem, William Hyde Wollaston reported his attempts to improve the dispersion of prisms so that they could spread the sunlight out into ever wider rainbows. Amid the laundry list of transparent substances that he had run sunshine through, he announced something odd. The continuous rainbow wasn't. Instead of an unbroken band of light, the sun's spectrum was cut into pieces by strange dark lines.

Wollaston naïvely suggested that these lines were, perhaps, the true dividing lines for the colors.

Prisms improved and scientists took note. By 1817, Joseph Fraunhofer had painstakingly catalogued dozens of these dark lines in the spectrum of the Sun and, for the first time, in the rainbows of the stars. His alphabetical labeling system for the most prominent lines is still in use. For instance, there are the D lines, two dark stripes in the orange-yellow region of the rainbow. Were you to hold up a DVD or CD at an angle to a low-pressure sodium street lamp, you would see not a rainbow but a distinct orange-yellow blob. And if you could measure the wavelength, you'd find that it fell at the same place as Fraunhofer's D lines. In 1817, though, nobody had any clue that sodium atoms have a special relationship with those particular colors.

Fraunhofer was also tantalized by the observation that *some* of the dark lines in the rainbows of *some* stars matched those in the Sun, but others did not. The strengths of some lines even depended on the time of day and the time of year, leading early spectral explorers (or spectroscopists) to the appropriate conclusion that some component of Earth's atmosphere was filtering out those colors. But the logical deduction that the other, unchanging lines must represent something in the makeup of the Sun and the stars had to wait half a century for the work of Gustav Kirchhoff and Robert Bunsen.

You have most likely heard of Bunsen, or at the least had experience with the gas burner that bears his name, despite the fact that he didn't invent it. Bunsen did, however, make a number of game-changing contributions to chemistry and geology for which he is scarcely remembered. Like most chemistry students, Bunsen was a fan of incinerating various substances in what is now known as a Bunsen burner. He was particularly surprised by the "splendid light"—a brilliant white light reminiscent of a welding torch—produced when a magnesium strip was inserted into the high-temperature flame. Spurred by a now obsessive drive to understand the light produced by burning various substances, Bunsen then found himself working with Kirchoff, who had "made a wonderful, entirely unexpected discovery in finding the cause of the dark lines in the solar spectrum . . . thus a means has been found to determine the composition of the Sun and fixed stars."

Perhaps it was a good thing that Comte died before Kirchoff and Bunsen got together to publish their findings. Now collectively known as Kirchoff's Laws of Spectroscopy (again Bunsen's contribution is ignored), they were the rules that the universe followed when creating light. The Sun's rainbow, including the dark lines cutting it up, was the natural result of running light from its hot, dense inner portion through the cooler, lower-density outer layers. The dark lines, far from being dividing lines for the colors, instead betrayed the identities of the various atoms making up that outer layer. In other words, the elements in the atmospheres of the Sun and stars were acting as filters, picking off the very same colors that they produced when Bunsen incinerated them. This meant that each element had its own spectral fingerprint, a set of wavelengths that it could either emit or absorb depending on the conditions.

Comte's certainty was shattered by equipment available to any modern high-school chemistry student. Light, even that from distant stars, could be unraveled and analyzed by the relatively simple spectroscope—a combination telescope, prism, and wavelength calibrator. Four decades after the work of Kirchhoff and Bunsen, Charles A. Young of Princeton University hailed the spectroscope as "a new and powerful instrument of astronomical research, resolving at a glance many problems which before had seemed to be absolutely inaccessible to investigation." Young even dared to suggest that "its invention has done almost as much for the advancement of astronomy as that of the telescope."

If nothing else, it filled the journals. Once the astronomical applications of spectroscopy came to light, the prestigious *Monthly Notices of the Royal Astronomical Society* frequently ran the equivalent of "how-to" articles in a very slow, decades-long version of a modern-day Internet chat. One letter in the 1864 volume of *MNRAS* even informed spectroscopic observers that an instrument capable of producing useful stellar spectra could be obtained for 66 Bavarian florins. So, it was a Craigslist, too, promoting the pots of gold at the end of the rainbow.

Then in 1880 a series of exceedingly practical articles by P. Smyth on the new field of spectroscopy appeared in the nascent journal *The Observatory*. The lexicon needed cleaning up. Standards needed to be adhered to. After two decades of independent inquiry with essentially homemade instruments, there was solid evidence that the spectroscope was indeed the new telescope. But understanding the Sun and stars would require all spectroscopists to be, well, on the same wavelength.

Over the next two decades Smyth's vision would be realized. In 1887, Norman Lockyer published *The Chemistry of the Sun*, then the definitive work on the subject, identifying the elements responsible for the Sun's dark-line, or absorption, spectrum. In 1895, Henry Rowland produced the first comprehensive solar atlas stemming from years of work at Johns Hopkins University. During the course of his research, Rowland obtained "the spectrum of every known element, except gallium (of which I have no specimen)." The final list—published in the very first volume of *The Astrophysical Journal*—contained 36 elements. Helium was added shortly afterward, when it was discovered in the spectrum of the Sun's chromosphere, a sizzling hot outer layer of the Sun that approaches 36,000°C (65,000°F) and that can be seen only during a solar eclipse.

31 THE SUN'S SECRET RECIPE

One major problem remained, though. How *much* of each element was present in the Sun and stars? Just knowing that it had calcium, for example, didn't help us know how abundant the element was. The most simplistic approach to getting the exact recipe for the Sun was to note which lines are the darkest. That would, in this naïve understanding, mean there had to be more absorbers picking off those particular wavelengths. More absorbers means darker stripes. Using Rowland's 1895 figures, calcium wins, with iron a close second. Hydrogen came in third place, but this was of little concern because most astronomers were then operating under the assumption that the composition of the Sun would mirror that of Earth. What concerned them was not that hydrogen should rank so low but that common terrestrial elements such as sulfur, nitrogen, and phosphorus did not appear on the list at all.

Others suggested that, instead of looking at the darkness of the stripes, they should simply count the number of spectral lines present. With this method, iron became the most abundant element, followed by nickel. This was a far more pleasing result, as Earth's density indicated a high proportion of these elements, and many out-of-this-world meteorites that had then been discovered contained primarily iron. By this reckoning, hydrogen ranked twenty-second on the list of the 36 known solar constituents. The rare-earth element yttrium ranked higher than the far more common element copper using either approach, and yet it's an element whose existence is completely unknown to the average person.

It was clear that either the interpretation of the spectra or the assumption about the solar composition was incorrect (or both). Although dead, Comte seemed to be constantly reminding the upstart astronomers of his certainty. Maybe we couldn't figure out the actual recipe for the Sun. Maybe we would have to content ourselves with getting the ingredient list.

But all was not lost. In a remarkable feat of foreshadowing, Norman Lockyer had suggested in 1887 that the dark lines tell us not about the actual recipes of the stars but instead about the conditions in the stars themselves:

> Lockyer thinks it more probable that the missing substances are not truly elementary, but are decomposed or dissociated by intense heat, and so cannot exist on the sun, but are replaced by their components. He maintains, in fact, that none of our so-called "elements" are really elementary, but that all are decomposable and are to some extent actually decomposed in the sun and stars.

Perhaps elements *can* be split apart. Perhaps the extreme conditions within stars render certain substances effectively invisible. But J. J. Thomson's discovery of the electron was ten years in the future, and even the most rudimentary understanding of the structure of the atom and the underlying mechanism behind the spectral lines would take yet another decade to materialize. Lockyer, despite being basically correct in his instinct, would never learn the secret recipe for the Sun. He died in 1920, five years before the Sun's main ingredient would be reported apologetically in what astronomy legend Henry Norris Russell judged "the best doctoral thesis I have ever read."

The author of that doctoral thesis was a woman named Cecilia Payne. Born in 1900, Payne grew up amid a thorough scientific revolution in a household where autographs of naturalist Charles Darwin and geologist Charles Lyell were among the prized family possessions. The classical physics that ruled until the end of the nineteenth century was giving way to a bizarre system that explained the mysteriously constant speed of light, the capricious behavior of the atom, and particles as waves. Geology was beginning to understand that *terra firma* is anything but, and biology was evolving before our eyes. Her undergraduate studies in Cambridge gave her a taste of all these subjects, but physics caught her attention, despite the outright bullying she received from physics giant Ernest Rutherford. His personal attacks, along with a well-timed talk by eminent astronomer Arthur Eddington, who had just led the pivotal expedition to witness the solar eclipse that verified Einstein's bizarre theory of relativity (see part I), ultimately drew her to astronomy instead. She earned her undergraduate degree at Cambridge but was unable to continue graduate studies there. She was, after all, a woman, and it would be another quarter of a century before Cambridge awarded higher degrees to anyone without a Y chromosome.

Luckily for Payne and for astronomy, on the other side of the pond, the Harvard College Observatory had recently set up a fellowship exclusively for female observers. Although Harvard, too, refused to confer advanced degrees on women, those could be taken care of by its sister institution, Radcliffe, a ten-minute walk up the road from Harvard. Payne soon found herself in a new Cambridge, this one in Massachusetts, involved with one of the most active astronomy research groups in the world.

In 1923, the HCO was under the directorship of Harlow Shapley, who had honored Payne's audacious request to work with some of the greatest astronomers of the day. She began work under the stellar classifier Annie Cannon, one of the few women whose names earn a place in introductory textbooks. Payne's intelligence and originality soon set her apart from the field, and she pursued her own research project. It was nothing short of ambitious: she would determine once and for all the exact recipe for the Sun. Just two years later, a day after her twenty-fifth birthday in May 1925, Payne became the first person—male or female—ever to earn a PhD in astronomy for work done at the Harvard College Observatory.

Her dissertation title was a pile of jargon hardly distinguishable from every other technical paper at the time: "Stellar Atmospheres: A Contribution to the Observational Study of High Temperature in the Reversing Layers of Stars." But her explanations within her dissertation were extraordinarily lucid. She spent the first 90 pages laying "The Physical Groundwork." With incredible clarity in writing that is almost extinct in today's dissertations, she artfully solved the problem of the variety of dark stripes in stellar rainbows by appealing not to the recipes themselves but rather to the temperature scales. The makeup of the stars, she found, is actually incredibly uniform. The difference lay mostly in their surface temperatures and hence the energies of light interacting with the elements.

To understand why the temperature makes such a difference, consider the simplest element hydrogen, which consists of a single proton and a single electron and typically no neutrons. Just a decade earlier, Niels Bohr had found that the electron seems to orbit the proton, somewhat in the fashion that a satellite orbits Earth. But just as launching a satellite with too much thrust will send it well outside Earth's gravitational influence, giving an electron too much energy will eject it from the proton's grasp. And, just as giving the satellite a bit more or a bit less energy will force it into a bigger or smaller orbit, giving an electron energy (or taking it away) might force it to be farther from or closer to its proton.

A hydrogen atom is assembled in such a way that light with ultraviolet wavelengths of 91.2 nanometers, equivalent to 91.2 billionths of a meter, or less can rip the electron from its parent proton. (Light is funny that way; the shorter the wavelength, the *greater* the energy.) The highest temperature stars—stars over about 9,700°C (17,500°F)— would be producing so much of this high-energy light that virtually all their hydrogen atoms would be just a frantic plasma of zippy protons and their long lost electrons. Producing the dark stripes that Wollaston first witnessed, however, required the electrons to still be orbiting their protons. From there, they could then absorb certain specific colors (for instance, a particular red wavelength of 656.3 billionths of a meter) only if there were enough atoms with electrons in the right state to pick off that color. That right state, Payne realized, depended on a very fine balance between light with enough energy to get them to their starting point and light with enough energy to get them to their endpoint but not so much energy that it ejected them entirely from the atom. However, the interiors of stars emit a whole rainbow of wavelengths, even many we can't see. For cooler stars, the rainbow is heavy on the lower-energy reds and oranges; for the hottest stars, it's heavy on the higher-energy blues and violets and even ultraviolet. Each atom was a miniature Goldilocks looking for a "just right" balance of energies. That balance could be achieved only with a "just right" temperature, and the "just right" temperature that allowed one element to express itself through absorbing certain wavelengths was different from the "just right" temperature of another element. Thus two objects can have precisely the same chemical makeup but have vastly different dark lines in their rainbows because their light balances—and hence, temperatures—were different.

Of course. It made so much sense. Now that the interaction of light with matter was better understood, the problems of both stellar composition and stellar temperature were elegantly solved. Payne even explained the apparent absence of the common elements oxygen, sulfur, and nitrogen in the Sun. It seems that the spectroscopist's usual range of observed wavelengths did not include the favorite wavelengths of these elements. Oxygen preferred to pick off light in the infrared, a type of light with wavelengths longer than red. Payne optimistically reported that oxygen determinations "should prove accessible in the near future." (Annoyed spectroscopists 90 years later still wrestle with figuring out the exact amount of oxygen in stars.)

It took a century of work and more than a dose of incredible intuition, but the Sun's recipe was finally within our grasp. Even better, everyone could now get back to the happy knowledge that the entire universe—including Earth—was utterly uniform in its makeup.

Or was it? Buried within her eloquent manuscript was a chapter where she discussed the proportions of ingredients in the Sun. It seemed that the simplest atoms—hydrogen and helium—were proving to be troublesomely dominant. But, despite the fact that atomic physicists understood more about the behavior of these two atoms than the rest of the denizens of the periodic chart, Payne cautioned that "the enormous abundance derived for these elements in the stellar atmosphere is almost certainly not real." On top of that, stars seemed positively anemic. Their iron content appeared to be nothing like that of Earth.

But why should she have doubted her results? She had applied the physics masterfully throughout her thesis. The result, even if it made no sense to her or to anyone else, was still the result. Right? Isn't that the way science progresses?

Sadly, even in the sciences, there is often pressure to conform. In Payne's case, the pressure came from scientific giant Henry Norris Russell, a professor at Princeton who was generally accepted to be the authority on all things astronomical. (He later literally wrote the book on astronomy—with two coauthors, that is—that would be the standard for a generation.) She wrote in a letter to a friend that, although she had overcome her personal fear of him, "His power in the astronomical world is another matter, and I shall fear that to my dying day, as the fate of such as I could be sealed by him with a word." After examining her results, Russell wrote a few helpful hints on ways to get the recipes for the Sun and Earth to better agree with each other. Then the hammer fell. "There remains one very much more serious discrepancy," he wrote. "It is clearly impossible that hydrogen should be a million times more abundant than the metals."

A footnote in Payne's dissertation indicates that Russell was in favor of reducing the amount of hydrogen and of raising the bar for iron "by a factor of at least 3 and probably 5—which would put it where it obviously belongs."

"Clearly impossible." "Obviously belongs." Phrases like these are hard to dispute when you are a lowly PhD student, particularly a female one in the early years of the twentieth century. Payne's results had to grow on the scientific community. Frustratingly, it was Russell himself who just four years later would report on the high hydrogen abundance in the Sun. His recipe for the Sun called for one and a quarter cups of hydrogen, a teaspoon of helium, a teaspoon of oxygen, half a teaspoon of "metallic vapors," and a dash of free electrons, mixed well and baked at 10,000°F (recipe may be doubled, tripled, or multiplied by a factor of a billion billion trillion as necessary). In his discussion of the "non-metals," a singularly uninformative term that astronomers use to describe only the two lightest elements, hydrogen and helium, Russell mentions that the fact that visible hydrogen lines show up at all in the Sun is nothing short of a miracle. You see, any electrons that are forced by the right energy light to get into the correct starting level will virtually instantaneously head somewhere else before they get a chance to pick off their favorite colors. Getting enough hydrogen atoms in the right state to filter out what we witness as the dark hydrogen lines is a bit like herding countless tiny negative cats. To get just a few to behave means that you must start out with an insane number of them.

"The obvious explanation," Russell stated casually, "—that hydrogen is far more abundant than the other elements—appears to be the only one." Although Russell made a brief mention of her thesis in his comparison with previous results, that comment was buried 53 pages into a 71-page journal article. The astronomical community assumed the breakthrough was his.

Payne went to her grave in 1979, having claimed for 50 years that the most important thing about Nature is that we continue to make discoveries about it, not that any particular person is given credit. It's not as though her accomplishments went unrecognized. Her career was replete with accolades and honors, including a lifetime achievement award, which was, ironically, the Henry Norris Russell Lectureship. As for her early dread of him while still a graduate student, Payne had her own fearsome reputation. She was known to have a wicked temper lurking behind an often gracious and open demeanor, and she scared at least one student into another field by smashing a collection of glass photographic plates to her office floor in rage. Others reveled in her "delightful chaos." The astronomical community collectively reveres her, though, for so clearly lighting the way to understanding the makeup of the Sun and stars.

33 THROUGH A STAR DARKLY

In the 1920s, the Sun's exact recipe was still very much uncertain, and even Payne and Russell knew that. The values describing the favorite electron hangouts for many atoms were guesses at best, but, as our understanding of inconceivably small objects grew, so did our understanding of colossally large objects that could easily house millions of puny Earths. Spurring progress was the theoretical breakthrough that nuclear fusion—the merging of hydrogen nuclei to create helium nuclei—was the source of the Sun's immense energy, an idea that took hold in the early 1930s. To make the sunshine in which we bask, the Sun destroys four million tons of matter every single second, matter which is converted into energy, as described by Einstein's famous equation $E = mc^2$. That energy then has to work its way through all the layers that are trying to filter some of it out.

Although you can't tell it from here, the type of light created in the Sun's dense, seething furnace would kill us instantly. Every time the protons of once-whole hydrogen atoms are fused together into a helium nucleus in the dense inferno of the Sun's core, gamma rays are produced. Fortunately, the Sun's core is enshrouded by 700,000 kilometers (435,000 miles) of insulation. From the instant it is made to the instant that it finally shines forth in splendor from the solar surface, the energy created in the nuclear reactions endures a random bumper-car ride that lasts anywhere from 17,000 to 100,000 years (some calculations put it as high as one million years; in any event, the sunshine you see today was actually made in the Sun's core sometime in the Pleistocene epoch). During this time, the unimaginably short wavelength gamma rays gradually donate their energy to electrons, protons, and various atoms that have had any number of electrons stripped from them. By the time the energy finally elbows its way to the surface of our star, it is almost exclusively low-energy light, something earthly life-forms can actually make use of.

In this way, the Sun provides us with a sort of physical lesson for our constant tendency to ask "what good is it?" Certainly if we had created a committee to require the Sun to make copious amounts of visible, life-giving energy, the committee would hardly have suggested that the most reasonable first step is to create an incredible amount of something that is fatal to us. But that "useless" energy, especially the way in which it interacts with the two trillion quadrillion tons of material in its way, shapes the conditions for every point in the entire Sun. Only after thousands of years of small steps—many of them backward steps, at that—and complicated interactions, it emerges as the driver of all life on our planet. One wonders whether in a universal budget crunch, the cosmic politicians would look at the Sun as anything but a dynamic whole. If so, would they see the comparatively tiny fusion reactor as a waste of resources that create nothing but a harmful product? It seems ridiculous, yet human society seems all too adept at compartmentalizing its own endeavors while failing to see the power of all the random walks.

At this point, though, the random walk of understanding the Sun's makeup seems a long way from improving anyone's cancer treatment.

34   THE OPACITY PROJECT

It is impossible to imagine exactly how the light and matter dance with and wrestle each other at every layer of the Sun during sunshine's long trek outward. All that energy causes a sizeable fraction of the Sun's interior to churn like a lava lamp as it strains to push out the energy of seven trillion Hiroshima blasts every second. The rest of the time the particles play lacrosse with the light, throwing it from one thing to another. A photon, or piece of light, can manage to go all of about a centimeter at a time before slamming into a new obstacle. Unfortunately for solar astronomers, understanding the precise interactions at every step of the way is crucial to understanding the Sun's exact makeup.

The fact that the Sun is practically next door would seem to make figuring out its exact composition child's play. On top of the Sun's proximity, there also is an entire solar system of objects that formed from the same cloud some 4.6 billion years ago to use for comparison's sake. If meteorites contain, say, two parts cadmium for a million parts silicon, it makes sense that the Sun should, as well. For decades, solar astronomers and meteorite hunters tried in vain to get all of their answers to mesh, but despite advances in every technology available to them, everyone finally had to agree to disagree.

It was, quite honestly, an embarrassment.

In 1984, chagrined by the fact that we still hadn't figured out the composition of our own star, Michael Seaton launched the Opacity Project. Sporting white Isaac Asimov sideburns and a mop of dark hair, Seaton had a talent for understanding not only the complex dances between particles and light but also the even more complex human interactions between scientists.

Seaton was always immensely supportive of his colleagues and students. In 1966, when his graduate student R. J. Harman died tragically in an accident, Seaton told Harman's parents that he would be remembered for his important astronomical work. To this day, one of the fundamental relationships in the study of the remnants of sunlike stars is called the Harman-Seaton sequence. It was this sort of human touch, along with a hard-working, positive attitude, that strengthened countless connections across disciplines. By the time he started the Opacity Project, Seaton had already spent more than three decades attempting to quantify the complicated choreography of atoms, all the while earning the support and admiration of his peers. In fact, just prior to creating his all-star team of physicists, astrophysicists, and programming experts, he was elected president of the British Royal Astronomical Society.

The random walk that started nearly two centuries earlier with an attempt to make an ever-wider rainbow had led to this perfect confluence of personality and technology. Former student and colleague Helen Mason recalls, "The decade of the 1980s was a golden age for atomic physics and spectroscopy. Mike Seaton was in the front row." Seaton would eventually assemble a team of more than 30 astronomers and physicists from France, Germany, the United Kingdom, the United States, and Venezuela. Their charge? Compute how strongly different wavelengths of light interact with different atoms. This yields something called the opacity, or the tendency for the light to be blocked by the atoms. To do this for all the atoms at all the layers of the Sun would require computers with more horsepower than ever. Their first results would take nearly a decade to compile.

It might seem on the face of it like an awful lot of trouble to go through simply to get the Sun's exact recipe. But it wasn't just the Sun's composition that was a problem. The Sun and other stars have a constant flurry of seismic waves traveling through them and across their surfaces. Just as earthly seismic waves help us to divine the inner structure of our planet, these pressure waves in the Sun give us information about the Sun's interior. The interior wasn't shaping up as it should, though, at least not if we understood how the light and atoms in the Sun were behaving. Many answers were distressingly uncertain. The amount of carbon, nitrogen, and oxygen in the Sun had to be adjusted by as much as 45 percent to get models and observations to mesh. This sort of discrepancy threw our understanding of the Sun, along with huge areas of astronomy, into question.

"The practical necessity of solving this problem can hardly be overstated," reads a line in a paper authored by many members of the Opacity Project. Sure, for astrophysicists, but what about the rest of humanity?

35   THE IRON LADY AND THE GOLD STANDARD

You might recall from science class that atoms consist of protons, neutrons, and electrons. The protons and neutrons reside together in the nucleus of an atom, while the electrons circle around them comparatively far away in fixed orbits. A scale model of a hydrogen atom would have a one-proton nucleus several meters wide, and a puny, few-millimeter electron many kilometers away. Hydrogen is simple, though, with only one electron to manipulate. The energies required to shuffle its electron around, or even rip it from its nucleus entirely, are well understood; the closer the electron is to the nucleus, the more energy is required to pull it out of its orbit. The problem of opacity—the amount of light absorption by the material in the Sun—does not arise with the abundant stores of hydrogen that the Sun and stars possess. It arises with the other elements. Something like iron, for instance, has 26 protons and, under circumstances that you've personally encountered, 26 electrons.

The interior of the Sun is unlike anything you've encountered, though, so some atoms will have had one, two, or even thirteen electrons ripped from them depending on where they are in the Sun. An iron atom hanging onto all of its electrons interacts with its own favorite wavelengths of light, whereas an iron atom with only 15 of its electrons interacts with a completely different set of wavelengths.

At Ohio State University, Sultana Nahar has paved the way in computing the specific interactions between light and, in particular, iron. A woman who claims she "never missed a math problem. Never," during her elementary and high school days in Bangladesh, Nahar quickly gained the reputation as the Iron Lady of Mathematical Computations. In 1987, she earned her PhD in atomic physics and soon found herself playing a vital role in Seaton's Opacity Project and its offshoot, the Iron Project. Her mathematical models of the detailed interaction between high-energy light and atoms, aided by the Ohio Supercomputer Center, have helped reduce—but not eliminate—the uncertainties about our own star.

The insides of stars aren't the only places where atoms are subjected to enormous energies, though. Regions around unimaginably dense black holes (see part III) are also flooded with high-energy light. There the atoms make up a violent plasma hurricane as they spiral toward their doom in the black hole. Other cosmic maelstroms include the leftovers from supernova blasts, the catastrophic explosions of stars, and even the dynamic centers of young galaxies. Knowing how the matter and energy theoretically interact in such conditions has helped reconcile observations of these remote, energetic objects. Nahar's computational expertise has reached into the Sun and across the known universe.

By the time Nahar joined the Opacity Project in 1990, her colleague Anil Pradhan at Ohio State University was already a seasoned Seaton veteran. With fellow Opacity Project physicist Yan Yu of Thomas Jefferson University, the three made a formidable team that was awarded hundreds of thousands of dollars in grants from American taxpayers through NASA, the National Science Foundation, and even the U.S. Department of Energy. They explored the interaction of light and matter, digging deep and finding sometimes surprising results.

As scientists had known since the work of Lise Meitner in 1922, the right type of light aimed at an atom can pull one of the atom's inner electrons out. Normally, scientists think of only the outer electrons interacting with light and with each other. When that happens, the other electrons will adjust accordingly. But pulling an inner electron out is kind of like hollowing out the foundation for the other electrons.

What Meitner found out as she explored radioactivity is that electrons, particularly those from the orbit immediately above the hollow, will tend to want to head "inward" to fill that space. But each time an electron falls into an orbit closer to the nucleus, it emits a certain amount of energy in the form of a photon. This photon can then knock one or more other electrons, leaving still more electron vacancies in the process. Other electrons from upper levels will continue to fall to fill the vacancies, each time releasing another photon. Just as removing a block of jewels or candy pieces from the bottom of popular video games causes the other pieces to slide down, sometimes causing unanticipated (and high-scoring) chain reactions, pulling an inner electron from an atom can result in a series of rearrangements and chain reactions of light and matter. Since the innermost electrons are the ones held most tightly to the nucleus, pulling one of those out requires the atom to absorb enormous, but very specific, energies—x-rays or gamma rays—and the subsequent rearranging of the leftover electrons can result in the atom spitting out things you wouldn't normally expect. Ripping out one of the innermost electrons in a gold atom, for instance, can result in the forceful expulsion of many energetic electrons. One parcel of the correct-energy light comes in, but lots of electrons—as many as 20—leave. This property is now known as the Auger effect, named after Pierre Auger (o-zhay, not aw-gher, although the electrons seem to be augering in), who quantified the behavior more precisely in 1923.

Astronomically speaking, this situation explains quite a bit. At certain so-called resonance frequencies, the gold atoms will tend to absorb light energy very efficiently. Far from being an effect that astronomers can simply average over, as they had largely done in the past, it profoundly changed the steps of the light-and-matter dance in extreme environments. When a single packet of light interacting with a heavy element can result in the sudden release of a dozen or more electrons, each of which interacts with light in its own way, even tiny amounts of the heavy elements can make their mark. Such resonances accounted for quite a bit in our understanding of opacity.

And thus the Opacity Project and the Iron Project progressed, incrementally adding to our understanding of interactions on unfathomably small scales and how those interactions affect our observations of phenomena on unfathomably large scales. Yan Yu dropped off Seaton's team in 2003, having devoted two decades of his life to the project, and went to join the ranks of medical physicists. Despite changing fields, Yu kept in close contact with his former colleagues and friends, Pradhan and Nahar. During casual conversation with Yu, Pradhan discussed the discovery of particular resonances of x-rays with heavy atoms. Had he been talking to just another atomic physicist, the exchange might not have gone anywhere. But Yu was now the director of Jefferson Medical College's Division of Medical Physics, an expert on radiation cancer therapies, and someone who could see such a breakthrough from a different angle. The resulting brainstorming session took a turn not to the hearts of atoms or the insides of stars but to people.

Seaton himself died in 2007 at the age of 84, his obituary penned by Pradhan and Nahar, who stated, "There are precious few scientists who have his unique abilities that ranged from profound theoretical insights to mathematical formulations and highly technical computational developments. Mike Seaton was an immense source of inspiration to all who knew him."

36   THERAPY OF THE STARS

If you go to the doctor to get an x-ray, you are voluntarily exposing yourself to a type of high-energy light that can literally pull molecules apart and rip electrons from their parent atoms. The trade-off is fairly clear, though. You will also get the chance to see whether that pain you've been feeling is the result of a fracture or, perhaps, something worse. The typical machine used to image your insides, either a standard x-ray machine or a CT scanner, produces a fairly wide range of energies, all the way from low-energy x-rays that are stopped pretty effectively by your body up to high-energy x-rays that pass through you without interacting strongly with anything. The medium-energy x-rays are often the most useful, interacting with your tissue and bone in such a way that important information can be gleaned from the images.

X-rays are also part of routine cancer therapies. These high-energy types of light damage the DNA inside cells, preventing the cells from dividing. Given that x-rays and other types of radiation are not particular about which types of cellular DNA they destroy, oncologists try to localize the energy around the site of the cancer. In some cases this involves placing a radioactive substance that spits out high-energy light or fast-moving charged particles inside the body itself (as in the case of Henrietta Lacks), while in others, a beam of radiation is aimed at the area containing the cancer. In the former case, the patient is actually radioactive until the substance is removed, forcing the patient to limit contact with others, particularly children. In the latter case, sometimes a daily radiation treatment is required for weeks on end in an attempt to destroy (or at the very least, diminish) the tumor.

It's not difficult to see why this type of treatment can have serious physical and emotional side effects. Radiation aimed at a tumor will kill not just the cancer cells but very likely also the healthy cells nearby. It is not unusual for cancer patients to feel exhausted and nauseated and to suffer great pain. Down the road, the radiation used to kill the cancer might itself be the cause of new cancerous growths.

Despite all of our advances in medicine, our methods for dealing with cancer can sometimes seem strangely barbaric. And since cancer is estimated to affect half the men and a third of the women in the United States at some point in their lives, everyone's life is impacted by it. What could be a higher priority for research funding than curing this insidious disease that will touch us all? Indeed, a 2013 survey indicated that five out of six Americans supported increased funding for cancer research, making it a national health priority. Unfortunately, budget woes resulted in a sequestration that reduced the National Institutes of Health budget by 5 percent in 2013. Although NIH scientists were hopeful that 2014 would see a rebound, funding remained disappointingly flat.

But this is where that two-century hunt for the Sun's recipe can come into play. Nahar, Pradhan, and Yu proposed something called resonant nano-plasma theranostics (RNPT) for a more precise treatment of cancer. They, and ultimately a host of other collaborators, came to realize that if gold or platinum nanoparticles ("nano" indicates the tiniest of workable scales) could be delivered to a tumor, the Auger effect that was discovered to play such a vital role in our understanding of the Sun's opacity and the extreme conditions around black holes could be exploited. The burgeoning field of nano-biotechnology is at work creating what would amount to chemical Velcro that attaches to cancerous tumors. Should these nanoparticles then be subjected to relatively low-energy x-rays tuned to one of those special resonance frequencies, the Auger effect will pull the inner electrons out, resulting in the flood of electrons into the tumor itself.

One of the great advantages of this idea is that those monochromatic x-rays would interact quite strongly with the gold or platinum nanoparticles but much less so with the patient's own tissue. The other advantage is that the energetic electrons released would flood only the sites of the nanoparticles, which have embedded themselves in the cancerous tumor itself. Those energetic electrons would then destroy only the tissue in closest proximity to those particles, namely the tumor. The unlikely band of over a dozen astronomers, physicists, biophysicists, and oncologists estimate that the treatment times and radiation dosages could be cut down by a factor of 10 or 100, certainly a welcome victory for patients currently looking at starting their radiation therapy regimens.

Sad to say, such a specially tuned x-ray device is not as easily obtained as the typical x-ray machine in your local hospital. Only a few such facilities exist in the world, largely in major physics laboratories where medical research applications have only just begun warming up. The RNPT team had an idea of devising a table-top monochromatic source by using the fluorescent properties of atoms, and on paper it is a sound design. However, more work is needed to develop a machine that can generate these x-rays with higher intensity. Furthermore, the required nanoparticles have not been fully developed. For now, the idea of RNPT looks spectacular as a concept, the theory and application fully hashed out by the considerable group of experts, but the practice has been frustrated by funding roadblocks. Feeling the sting of budget cuts, NIH is hesitant to provide support for an as-yet-hypothetical treatment, despite the promise.

Still, the team is forging ahead, with one member spending six months at a European particle accelerator facility to validate the methodology using monochromatic x-rays. Meanwhile, Ohio State University Medical Center's Rolf Barth has added his expertise on platinum compounds used in chemotherapy, as well as laboratory facilities to perform in vitro and in vivo experiments, both of which are practically mandatory for the NIH to be fully persuaded. Preliminary trials on cancerous cells have proven to be promising, and the team is in the process of translating their findings to experiments on live mice. While those experiments have not employed the monochromatic x-rays envisioned, their results are giving the team some hard data for the grant reviewers to consider.

Two decades ago, neither Nahar nor Pradhan would have predicted this career trajectory. Happily immersed in computational physics, it never occurred to Nahar that her work might impact lives in this fashion. She was simply doing what she loved doing, and that was to understand how light and matter affect each other at the most fundamental level. Should this treatment see the light of day, it will owe itself in large part to her curiosity about those minute interactions. In fact, without that basic curiosity, such an idea would never have even made it to the drawing board. "Oncologists *do* have physics backgrounds," she argues. "But they are not involved with detailed research of photon-atom interaction."

Pradhan, ever optimistic about the chances of RNPT to make it through the funding hoops, explains further: "Our work exemplifies fundamental science and the underlying symbiosis between apparently disparate branches of science. This is where new ideas arise."

✳ ✳ ✳   Now imagine. In a budgetary spiral, funding is squeezed from NASA, from the NSF, from fundamental research questions in favor of immediately practical applications. RNPT never sees the light of day, nor do dozens of other fruits of the curiosity and symbiosis that define science. Imagine stifling creative connections that could have been made in the labs of scientists who aren't in the business of battling cancer, whose work has no promise of immediate practical benefit. Imagine that your daughter's cancer treatment mirrors that of Henrietta Lacks all those decades ago, all because questions about the stars seemed too remote for the taxpayers to fund.

AFTERWORD

Researching this book has required countless hours of reading about scientific fields I'd never before explored and learning about scientists and discoveries that, to my shame, I'd never even once heard of. Entire partially written chapters were planned around profound connections that have been made, but not fully realized, or smaller connections that have been fully realized—even patented—but whose histories didn't seem as rich and intertwined as the five I ultimately chose to include.

On an almost daily basis, a new article describing a connection between some esoteric scientific pursuit and a tangible benefit showed up on my Facebook feed (I make no apologies for having "liked" so many science groups). Even in my pleasure reading, I could not escape the theme of this book. As I finished writing the final chapters of this book, I read Marcus du Sautoy's *The Music of the Primes* (I make no apologies for reading a book about mathematics in my spare time, either). I became fascinated by the link between what had to be seen as one of the least practical of all human endeavors—exploring the series of prime numbers—and the electronic encryption that keeps everybody's money secure. What could be more life-changing than every electronic monetary transaction made, each of which rests tenuously on our mathematical ignorance of patterns in the prime numbers? How prime number theory evolved to become the only barrier between hackers and your credit card information almost became its own section of this book (I even had a catchy title: "A Prime Example"), but I couldn't imagine that my version would have been anything more than a simplified retelling of du Sautoy's already thorough treatment.

What I noticed about all the ideas that vied to become part of this book is a common theme: the players in these stories had the ability to see vital connections where there seemed to be none. This sort of insight is a hallmark of creativity, the kind of thinking at which young children excel but that is slowly, methodically "educated" out of them as they seek to fill in the one and only correct answer to each question on a standardized test. Creativity expert and education reformer Sir Ken Robinson argues that true intelligence comes from decompartmentalizing subjects and seeing things more holistically. A paper clip is not simply a tool to clip paper together; it could also be a thousand other things. If only we could let go of the deeply ingrained notion that each tool—whether physical or mental—has a specific purpose, that there is one right answer, we would open up rich avenues of discovery. In a 2006 TED talk, Robinson stated that "creativity . . . more often than not comes about through the interaction of different disciplinary ways of seeing things." This sentiment was echoed by virtually every scientist I interviewed and in virtually every autobiography I read. Their ability to see connections spurred the creativity that would change lives.

But at what cost? True creativity in any field, science included, sees at least as many failures as success stories. Beethoven, for instance, wrote more than 200 pieces of music that he never bothered to include in his list of works because they either failed to meet up to his own high standards or remained unfinished. Works that he did finish, he started, stopped, rewrote, stopped again, and revised until he felt they were ready. He even wrote a number of musical jokes, possibly a result of tinkering around while searching for his muse. Given that he was essentially living on government money (he was granted an annual stipend from various members of the Vienna aristocracy), one might have argued that musical joking was a waste of funding and talent. Cynics might claim that if only he'd been more focused on creating masterpieces, he would not have spent so much time revising and discarding his work.

Yet his is a consistent pattern of creativity in composers, writers, inventors, artists, mathematicians, scientists, and even in Nature itself, where creatively branching into a new species often leads to an evolutionary dead end. Entire pathways are often discarded, but the creation of those can lead to new connections and insights. Failure is not only an option; it's apparently a requirement, but only if it becomes a learning experience and not a badge of inadequacy.

Beyond simply knowing that most creative endeavors will go nowhere, we must realize the pointlessness of trying to *force* someone to write a masterpiece or to see a connection they've never seen before. A casual observer might think the off-color limerick dashed out while the writer stares at the blank screen serves absolutely no purpose, but it might be just the thing to get the thoughts rolling again. A walk around the lake would seem to be an unproductive escape, except when it results in something like the "Eureka!" moment that sparked Einstein's relativity theory. And placing shrimp on a treadmill might ultimately tell us nothing whatsoever about anything practical. Or it might hold the key to salvaging a failing ecosystem. If we want those crucial connections to be made, the cost of encouraging creative minds to explore our world is fairly minimal.

For many, though, it isn't the monetary cost that concerns them. What worries them is the unintended dark side. The same thinking that led to GPS also led to $E = mc^2$, which provided the world with the atomic bomb and the Chernobyl disaster. And the GPS that saves the lives of stranded hikers also can cost the life of the unfortunate target of a determined and technologically savvy stalker. The genetic understanding that helps exonerate prisoners and pinpoint the roots of some diseases also neatly paves the way for the insidious practice of eugenics, where "bad" genes are weeded out of the population. Should a particular gene be implicated in a politically or socially unpopular trait, what repercussions would that discovery have? Sure, technological advances have created the convenience of cars and air conditioners, but those same conveniences have wreaked havoc with our climate and protective ozone layer. All things considered, many argue, wouldn't we be better off without all this curiosity-driven research? It seems that when an unintended benefit helps us make headway on one front (e.g., antibiotics), an unintended detriment (e.g., multiple-antibiotic-resistant bacteria) isn't far behind.

I maintain that it is precisely because of this potential for harm that people need to be educated like never before. And I don't mean educated to fill in the correct bubble on a standardized test but to see the big picture more analytically. As Carl Sagan pointed out, "We live in a society exquisitely dependent on science and technology, in which hardly anyone knows anything about science and technology. This is a clear prescription for disaster."

Part of the problem is that scientists aren't effectively communicating how their work fits into the grand scheme of things, but another part of the problem is that people often can't even *see* a grand scheme of things anymore. Parents, schools, and society are failing to foster the creative connections that will help future generations make the leap from a discovery in one field to an application in another. Even in many preschools, Play-Doh time is distinctly separate from story time, which is distinctly separate from music time, even though modeling a musical story would encourage the multidisciplinary thinking required to get us through the next decades.

Cutting back on curiosity is like silencing every child's "Why?" Yes, you will save yourself a few moments of time (possibly the embarrassment of being reduced to "I don't know."), just as cutting funding for curiosity will save perhaps a penny on every tax dollar, but at what cost? Curiosity helps humanity navigate the rich landscape of Nature, and every new discovery provides us with another piece of the map. We can use the map to aid in conquering and destruction, or we can use it to find an unexpected and scenic shortcut. Our obligation is not to destroy the map, to pretend that the landscape doesn't exist, or even to create an easier, false map, but to use the map wisely. This means encouraging intellectual cartographers when we can, but it also means learning more about universal mapmaking ourselves.

After all, this map doesn't come with a computerized voice to help us get to our desired destination. That's our job.

REFERENCES

Preface

Gaynes, D. 2012. "Saving Hubble." http://www.savinghubble.com/.

Progress on the James Webb Space Telescope found at http://www.nasa.gov /press/2014/february/nasa-administrator-bolden-senator-mikulski-view -progress-on-james-webb-space/#.UvJrSbQqrGo.

Introduction

American Association for the Advancement of Science Intersociety Working Group. 2013. *AAAS Report XXXVIII Research and Development FY 2014*. Washington, DC.

Average life span of a company found at http://www.businessweek.com /chapter/degeus.htm.

Cohen, I. B. 1987. "Faraday and Franklin's 'Newborn Baby.'" *Proceedings of the American Philosophical Society*, 131, no. 2, p. 177.

Football stadium construction cost estimates found at http://prod.static.vi kings.clubs.nfl.com/assets/docs/stadium/DES-recent-nfl-stadiums.pdf.

Golden Fleece Awards found at http://www.wisconsinhistory.org/turning points/search.asp?id=1742.

Golden Goose Awards found at http://www.goldengooseaward.org/.

James Webb Space Telescope statistics found at http://www.jwst.nasa.gov /faq_scientists.html.

NASA 2013 budget found at http://www.nasa.gov/news/budget/index.html.

National Science Foundation budget found at http://www.nsf.gov/about /budget/fy2013/.

NEA 2012 budget found at http://www.nea.gov/news/news12/Budget.html.

The Science Coalition: Sparking Economic Growth. 2010. http://www.sci encecoalition.org/reports/Sparking%20Economic%20Growth%20Full %20Report%20FINAL%204-5-10.pdf.

Tyson, N. D. 2011. *Real Time with Bill Maher* interview found at http://www .youtube.com/watch?v=3_F3pw5F_Pc.

U.S. funding for science estimate found at http://www.scientificamerican .com/article.cfm?id=money-for-science.

U.S. 2014 federal budget summary found at http://www.whitehouse.gov
    /sites/default/files/omb/budget/fy2014/assets/tables.pdf.

Part I. Finding Ourselves

"Atomic Clock to Check Einstein." 1959. *Popular Electronics*, vol. 11, no. 4,
    p. 66.
Beck, A., and Havas, P. (translators). 1989. *The Collected Papers of Albert Ein-
    stein*, vol. 2. Princeton University Press.
Chou, C. W., Hume, D. B., Rosenband, T., and Wineland, D. J. 2010. "Optical
    Clocks and Relativity." *Science*, vol. 329, no. 5999, pp. 1630–1633.
Coles, P. 2001. "Einstein, Eddington, and the 1919 Eclipse." *Historical Devel-
    opment of Modern Cosmology*, ASP Conference Series, vol. 252, pp. 21–41.
Crowell, B. 2009. *General Relativity*. Light and Matter, Fullerton, California.
Eclipse path for May 29, 1919, found at http://eclipse.gsfc.nasa.gov/SEhis
    tory/SEplot/SE1919May29T.pdf.
Global GPS market value estimate found at http://www.marketsandmarkets
    .com/Market-Reports/global-GPS-market-and-its-applications-142.html.
GPS life-saving benefits found at http://japandailypress.com/new-gps-net
    work-could-provide-faster-life-saving-tsunami-warnings-2029123.
"GPS Saves Boy's Life." http://www.cbsnews.com/video/watch/?id=2784982n.
"Gravity Probe B: Testing Einstein's Universe," found at http://einstein.stan
    ford.edu/index.html.
"How the Japan Earthquake Made the Day Shorter." 2011. *Popular Mechanics*.
    http://www.popularmechanics.com/science/environment/natural-disas
    ters/how-the-japan-earthquake-made-the-day-shorter.
Ives, H. E., and Stilwell, J. 1938. "An Experimental Study of the Rate of a
    Moving Atomic Clock." *Journal of the Optical Society of America*, vol. 28,
    no. 7, pp. 215–219.
Livingston, D. M. 1973. *The Master of Light: A Biography of Albert A. Michelson*.
    University of Chicago Press.
Lombardi, M. A. 2011. "The Evolution of Time Measurement. Part 3: Atomic
    Clocks." *IEEE Instrumentation and Measurement Magazine*, vol. 14, no. 6,
    pp. 46–49.
Lombardi, M. A., Heavner, T. P., and Jefferts, S. R. 2007. "NIST Primary Fre-
    quency Standards and the Realization of the SI Second." *NCSL Interna-
    tional Measure*, vol. 2, no. 4, pp. 74–89.
Lorentz, H. A., Einstein, A., Minkowski, H., and Weyl, H. 1952. *The Prin-
    ciple of Relativity: A Collection of Original Memoirs on the Special and General
    Theory of Relativity*. Dover, New York.
Matson, J. 2010. "How Time Flies: Ultraprecise Clock Rates Vary with Tiny
    Differences in Speed and Elevation." *Scientific American*, vol. 303, no. 3.
    http://www.scientificamerican.com/article/time-dilation/.

Maxwell, J. C. 1873. *A Treatise on Electricity and Magnetism*. MacMillan and Company, Publishers to the University of Oxford, London.

Pais, A. 1982. *Subtle Is the Lord: The Science and the Life of Albert Einstein*. Oxford University Press.

Ramsey, N. F. 1983. "The History of the Atomic Clock." *Journal of Research of the National Bureau of Standards*, vol. 88, no. 5.

Ramsey, N. F. 1993. "I. I. Rabi 1898–1988: A Biographical Memoir." National Academy of Sciences, Washington, DC.

Saha, M. N., and Bose, S. N. (translators). 1920. *The Principle of Relativity: Original Papers by A. Einstein and H. Minkowski,* with a historical introduction by P. C. Mahalanobis. http://archive.org/details/theprincipleofre00e insuoft.

Wick, G. 1972. "The Clock Paradox Resolved." *New Scientist*, vol. 53, no. 781, p. 261.

Part II. Identity Crisis

Allen, G. E. 2004. "Mendelian Genetics and Postgenomics: The Legacy for Today." *Ludus Vitalia*, vol. 12, no. 21, pp. 213–236.

Avery, O. T., MacLeod, C. M., and McCarty, M. 1944. "Studies on the Chemical Nature of the Substance Inducing Transformation of Pneumococcal Types." *Journal of Experimental Medicine,* vol. 79, no. 2, pp. 137–157.

Brock, T. D. 1995. "The Road to Yellowstone and Beyond." *Annual Reviews of Microbiology*, vol. 49, pp. 1–28.

Butler, J. M. 2010. *Fundamentals of Forensic DNA Typing*. Elsevier, Burlington, MA.

Dahm, R. 2008. "Discovering DNA: Friedrich Miescher and the Early Years of Nucleic Acid Research." *Human Genetics*, vol. 122, pp. 565–581.

deVries, H. 1903. "Fertilization and Hybridization." http://www.esp.org/books /devries/pangenesis/facsimile/contents/pangenesis-pp217-i.pdf.

Gosling, R., on Franklin/Wilkins found at http://www.dnalc.org/view/15261 -The-culture-in-Maurice-Wilkins-lab-Raymond-Gosling.html.

Griffith, F. 1928. "The Significance of Pneumococcal Types," *Journal of Hygiene*, vol. 27, no. 2, pp. 113–159.

Hayes, W. 1966. "Genetic Transformation: A Retrospective Appreciation. First Griffith Memorial Lecture." *Journal of General Microbiology*, vol. 45, pp. 385–397.

Henig, R. M. 2001. *The Monk in the Garden: The Lost and Found Genius of Gregor Mendel, Father of Genetics*. Mariner Books, New York.

The Innocence Project found at www.innocenceproject.org/.

Kornberg, A. 1989. *For the Love of Enzymes: The Odyssey of a Biochemist*. Harvard University Press.

Kresge, N., Simoni, R. D., and Hill, R. L. 2005. "Arthur Kornberg's Discovery of DNA Polymerase I." *Journal of Biological Chemistry*, vol. 280, no. e46.

Lehman, I. R. 2003. "Discovery of DNA Polymerase." *Journal of Biological Chemistry*. vol. 278, no. 37, pp. 33473–34738.

Mendel, G. 1866. "Versuche über Plflanzen-hybriden." Verhandlungen des naturforschenden Ver-eines in Brünn, Bd. IV für das Jahr 1865, Abhandlungen, 3–47; English translation by William Bateson, 1901. http://www.esp.org/foundations/genetics/classical/gm-65.pdf.

Mullis, K. 1990. "The Unusual Origin of the Polymerase Chain Reaction." *Scientific American*, vol. 262, no. 4, pp. 56–61, 64–65.

Mullis, K. 1998a. *Dancing Naked in the Mind Field*. Pantheon Books, New York.

Mullis, K. 1998b. Interview. http://www.nytimes.com/1998/09/15/science/scientist-at-work-kary-mullis-after-the-eureka-a-nobelist-drops-out.html?pagewanted=all&src=pm.

National Science Foundation history found at http://www.nsf.gov/news/special_reports/history-nsf/1950_truman.jsp.

O'Connor, C. 2008. "Isolating Hereditary Material: Frederick Griffith, Oswald Avery, Alfred Hershey, and Martha Chase." *Nature Education*, vol. 1, no. 1, p. 105.

Pray, L. 2008. "Discovery of DNA Structure and Function: Watson and Crick." *Nature Education*, vol. 1, no. 1, p. 100.

Rabinow, P. 1996. *Making PCR: A Story of Biotechnology*. University of Chicago Press.

Saiki, R. K., Gelfand, D. H., Stoffel, S., Scharf, S. J., Higuchi, R., Horn, G. T., Mullis, K. B., and Erlich, H. A. 1988. "Primer-directed Enzymatic Amplification of DNA with a Thermostable DNA Polymerase." *Science*, vol. 239, no. 4839, pp. 487–491.

Sanger, F., Nicklen, S., and Coulson, A. R. 1977. "DNA Sequencing with Chain-terminating Inhibitors." *Proceedings of the National Academy of Science*, vol. 74, no. 12, pp. 5463–5467.

Sutton, W. S. 1903. "The Chromosomes in Heredity." *Biological Bulletin*, vol. 4, pp. 231–251.

Upjohn company history found at http://www.fundinguniverse.com/company-histories/the-upjohn-company-history/.

Watson, J., and Crick, F. 1953. "Molecular Structure of Nucleic Acids: A Structure for Deoxyribose Nucleic Acid." *Nature*, vol. 171, pp. 737–738.

Yellowstone National Park information on bioprospecting and benefits sharing found at http://www.nps.gov/yell/planyourvisit/upload/247-8-10b.pdf.

Part III. Finding a Hot Spot

Alpher, R. A., Bethe, H., and Gamow, G. 1948. "The Origin of the Chemical Elements." *Physical Review*, vol. 73, no. 7, pp. 803–804.

Banks, M. 2010. "Launching the Wireless Revolution." *Physics World*, vol. 23, no. 7, pp. 12–13.

Cole, T. W. 1973. "Observations with an Electro-Optical Radio Spectrograph." *Astrophysical Letters*, vol. 15, pp. 59–60.

CSIRO budget information found at www.csiro.au/Portals/About-CSIRO/How-we-work/Budget—Performance/Portfolio-Budget-Statement.aspx.

CSIRO lawsuit information can be found at http://www.engadget.com/2006/11/16/csiro-wins-landmark-wlan-lawsuit-against-buffalo-more-to-come/.

Hankins, T. H., Campbell, D. B., Davis, M. M., Ferguson, D. C., Steber, W., Neidhofer, J., Wright, G. A. E., Ekers, R., and O'Sullivan, J. 1981. "Searches for the Radio Millipulses from the M87 Virgo A." *Astrophysical Journal Letters*, vol. 244, pp. L61–L64.

Hawking, S. W. 1971. "Gravitationally Collapsed Objects of Very Low Mass." *Monthly Notices of the Royal Astronomical Society*, vol. 152, pp. 75–78.

Hawking, S.W. 1974. "Black Hole Explosions?" *Nature*, vol. 248, pp. 30–31.

Hubble, E. 1929. "A Relation between Distance and Radial Velocity among Extragalactic Nebulae." *Proceedings of the National Academy of Sciences*, vol. 15, no. 3, pp. 168–173.

Nayak, B., and Singh, L. P. 2011. "Accretion, Primordial Black Holes and Standard Cosmology." *Pramana—Journal of Physics* (Indian Academy of Sciences), vol. 76, no. 1, pp 173–181.

Oppenheimer, J. R., and Snyder, H. 1939. "On Continued Gravitational Contraction." *Physical Review*, vol. 56, pp. 455–459.

Oppenheimer, J. R., and Volkoff, G. M. 1939. "On Massive Neutron Cores." *Physical Review*, vol. 55, pp. 374–381.

O'Sullivan, J. D., Ekers, R. D., and Shaver, P. A. 1978. "Limits on Cosmic Radio Bursts with Microsecond Time Scales." *Nature*, vol. 276, pp. 590–591.

Patterson, C. D., Ellingson, S. W., Martin, B. S., Deshpande, K., Simonetti, J. H., Kavic, M., and Kutchin, S. E. 2009. "Searching for Transient Pulses with the ETA Radio Telescope." *ACM Transactions on Reconfigurable Technology and Systems*, vol. 1, issue 4, article 20.

Penzias, A. A., and Wilson, R. W. 1965. "A Measurement of Excess Antenna Temperature at 4080 Mc/s." *Astrophysical Journal*, vol. 142, pp. 419–421.

Rees, M. J. 1974. "Black Holes." *The Observatory*, vol. 94, pp. 168–179.

Roguin, A. 2002. "Christian Johann Doppler: The Man behind the Effect." *The British Journal of Radiology*, vol. 75, pp. 615–619.

WiFi basic timeline of development found at http://wifinetnews.com/archives/2002/08/wi-fi_timeline.html.

WiFi saving lives example found at http://www.postbulletin.com/news/stories/display.php?id=1488756.

Part IV. Pick Your Poison

Azar, B. 2012. "QnAs with Baldomero Olivera." *Proceedings of the National Academy of Sciences of the United States of America*, vol. 109, no. 34.

Bourke, J. 2011. "The History of Medicine as the History of Pain." *History Today*, vol. 61, no. 4.

Burke, R. 2007. "Sir Charles Sherrington's 'The Integrative Action of the Nervous System': A Centenary Appreciation," *Brain: A Journal of Neurology*, vol. 130, no. 4, pp. 887–894.

Dafny, N. 1997. "Pain Principles," from *Neuroscience Online*. Department of Neurobiology and Anatomy, University of Texas Medical School at Houston. http://neuroscience.uth.tmc.edu/s2/chapter06.html.

Fein, A. 2012. *Nociceptors and the Perception of Pain*. University of Connecticut Health Center. http://cell.uchc.edu/pdf/fein/nociceptors_fein_2012.pdf.

Harrison, A. P., Hansen, S. H., and Bartels, E. M. 2012. "Transdermal Opioid Patches for Pain Treatment in Ancient Greece," *Pain Practice*, vol. 12, pp. 620–625.

Health research priorities of Americans survey found at http://www.fierce biotech.com/press-releases/phrma-launches-annual-exploration-ameri cas-health-views-concerns-and-progress.

Jack, D. B. 1997. "One Hundred Years of Aspirin." *The Lancet*, vol. 350, pp. 437–439.

Jeffreys, D. 2005. *Aspirin: The Remarkable Story of a Wonder Drug*. Bloomsbury Publishing, New York.

Machalek, A. Z. 2005. "Sea Snail Venom Yields Powerful New Painkiller." *The NIH Record*, vol. 17, no. 5. http://nihrecord.od.nih.gov/newsletters /2005/03_01_2005/story03.htm.

Mackowiak, P. A. 2000. "Brief History of Antipyretic Therapy." *Clinical Infections Diseases*, vol. 31, supple. 5, pp. S154–S156.

Mukhopadhyaya, R. 2012. "Classics: From DNA Enzymes to Cone Snail Venom: The Work of Baldomero M. Olivera." *Journal of Biological Chemistry*, vol. 287, no. 27, pp. 23020–23023.

Olivera, B. M., research group's website found at http://www.theconesnail .com/.

Olivera, B. M. 2006. "Conus Peptides" [three-part series of video lectures]. http://www.ibiochina.org/sem/BioChemPhy.htm.

Olivera, B. M., and Cruz, L. J. 2001. "Conotoxins in Retrospect," *Toxicon*, vol. 39, pp. 7–14.

Posadas, D. 2007. "Poison Pill." *Forbes*. http://www.forbes.com/global/2007 /0702/064.html.

Roach, J. 2005. "Toxic Snail Venoms Yielding New Painkillers, Drugs," *National Geographic*. http://news.nationalgeographic.com/news/2005/06/0614_05 0614_snaildrugs.html.

Stone, E. 1763. "An Account of the Success of the Bark of the Willow in the Cure of Agues. In a Letter to the Right Honourable George Earl of Macclesfield, President of R.S. from the Rev. Mr. Edmund Stone, of Chipping-Norton in Oxfordshire." *Philosophical Transactions of the Royal Society*, vol. 53, pp. 195–200.

Stufflebeam, R. 2008. "Neurons, Synapses, Action Potentials, and Neurotransmission." http://www.mind.ilstu.edu/curriculum/neurons_intro/neurons _intro.php.

Part V. Dreaming of the Star Treatment

Dronsfield, A., and Ellis, P. 2011. "Radium: A Key Element in Early Cancer Treatment." *Education in Chemistry*, vol. 48, no. 2, pp. 56–59.

Mason, H. E. 2008. "Emission Lines from the Solar Corona." *Astronomy and Geophysics*, vol. 49, no. 6, pp. 20–22.

Montenegro, M., Nahar, S. N., Pradhan, A. K., Huang, K., and Yu, Y. 2009. "Monte Carlo Simulations and Atomic Calculations for Auger Processes in Biomedical Nanotheranostics." *Journal of Physical Chemistry A*, vol. 113, pp. 12364–12369.

Nahar, S. N., Lim, S., Montenegro, M., Pradhan, A. K., Barth, R., Bell, E., Chowdhury, E., Turro, C., and Pitzer, R. M. 2012. "X-Ray Resonant Irradiation and High-Z Radiosensitization in Cancer Therapy Using Platinum Nano-Reagents." 67th International Symposium on Molecular Spectroscopy, Ohio State University, Columbus, Ohio, June 18–22, 2012.

Payne, C. H. 1925. "Stellar Atmospheres: A Contribution to the Observational Study of High Temperature in the Reversing Layers of Stars." PhD Thesis, Radcliffe College.

Pradhan, A., and Nahar, S. 2011. *Atomic Astrophysics and Spectroscopy*. Cambridge University Press.

Pradhan, A. K., Nahar, S. N., Montenegro, M., Yu, Y., Zhang, H. L., Sur, C., Mrozik, M, and Pitzer, M. M. 2009. "Resonant X-ray Enhancement of the Auger Effect in High-Z Atoms, Molecules, and Nanoparticles: Potential Biomedical Applications." *Journal of Physical Chemistry A*, vol. 113, pp. 12356–12363.

Radiation therapy for cancer information found at http://www.cancer.gov /cancertopics/factsheet/Therapy/radiation.

Russell, H. N. 1929. "On the Composition of the Sun's Atmosphere." *Astrophysical Journal*, vol. 70, pp. 11–82.

Storey, P., and Burke, P. 2008. "Mike Seaton's Legacy." *Astronomy and Geophysics*, vol. 49, no. 6, p. 15.

Tsamis, Y. 2008. "Integral Field Spectroscopy of Planetary Nebulae and Proplyds." *Astronomy and Geophysics*, vol. 49, no. 6, pp 16–17.

Wesson, R. 2008. "Planetary Nebulae, Novae, and the Abundance Discrepancy." *Astronomy and Geophysics*, vol. 49, no. 6, pp. 18–20.

Wollaston, W. H. 1802. "A Method of Examining Refractive and Dispersive Powers by Prismatic Reflection." *Philosophical Transactions of the Royal Society*, vol. 92, pp. 365–380.

Afterword

du Sautoy, M. 2003. *The Music of the Primes*. HarperCollins Publishers, New York.

Robinson, K. 2006. "Do Schools Kill Creativity?" TED2006 talk in Monterey, CA. http://www.ted.com/index.php/talks/view/id/66.

Sagan, C. 1990. "Why We Need to Understand Science." *Skeptical Inquirer*, vol. 14, no. 3.

INDEX

A41102 chip, 112, 115. *See also* fast
  Fourier transform; O'Sullivan,
  John
acetaminophen, 131
acetylsalicylic acid, 132–33. *See also*
  aspirin; Bayer (company); willow
adenine, 55, 69. *See also* DNA
agues, 128–29
α-conotoxin G1, 143
Alpher, Ralph, 100–101
American Astronomical Society, ix
American Institute of Mathematics, 8
American Type Culture Collection,
  81–82, 89
animal spirits, 135
Antifebrin, 131
Arecibo, 106, 110
aspirin: biochemical mechanism, 134;
  development of, 128–33. *See also*
  acetylsalicylic acid; Bayer (company);
  prostaglandins; willow
*Aspirin: The Remarkable Story of a Wonder
  Drug* (Jeffreys), 131
Astbury, William, 66. *See also* DNA
atomic clock: ammonia clock, 36; cesium
  clock, 37–38; in GPS, 14, 42–43;
  original concept, 16–17; precursor,
  34–35; as test of relativity, 38, 41–42.
  See also Hafele-Keating experiment;
  Mr. Clock; Rabi, Israel Isaac (Izzy);
  Ramsey, Norman
Auger, Pierre, 172. *See also* Auger effect
Auger effect, 172, 175. *See also* Meitner,
  Lise; opacity
Austek Microsystems, 112

Australian Square Kilometer Array Pathfinder (ASKAP), 119
Avery, Oswalt, 64–66

background radiation. *See* CMB
Barth, Rolf, 176
Bass, Lawrence Wade, 64
Bayer (company), 130–33
Bayer, Friedrich, 130–31
Beethoven, Ludwig van, 180
Bethe, Hans, 100
big bang, 100
black hole: detection of, 96; evapora-
  tion of, 104–5, 106; and massive star
  core collapse, 96; and opacity, 171;
  prediction of, 95; primordial, 103–7;
  supermassive, 96–97
Bohr, Niels, 160
Brock, Thomas: and Brock and Freeze
  YT-1 strain, 81–82, 89; discovery of
  Thermus aquaticus, 78–79; early life,
  75–76; work at Yellowstone, 77–79
Brock and Freeze YT-1 strain, 81–82, 89.
  See also Brock, Thomas; Yellowstone
  National Park
budget: NASA, 3; National Endowment
  for the Arts, 3; National Science
  Foundation, 3; United States, 3
Buffalo Technology, 117
Bunsen, Robert, 154–55

calcium ion channel, 144, 148–49
Caldobacter trichogenes, 78
Canadian National Research Council, 3
canal rays, 32